A COSMIC
CONSPIRACY

A COSMIC CONSPIRACY

Quantum Physics, Consciousness, and Secrets of the Universe

B.A. Crisp

2Portal Publishing

The publisher is not responsible for websites (or their content) that are not owned by the publisher. 2 Portal Publishing provides this author for speaking engagements. To learn more, please email: dcrisp@2portalpublishing.com

Library of Congress Cataloging—in—Publication Data

Crisp, BA 1966- A Cosmic Conspiracy: *Quantum Physics, Consciousness, and Secrets of the Universe*
—First Edition p. 510 cm.—(A Cosmic Truths Book)
ISBN (979-8-9925189-0-0 paperback) (978-1-7343087-9-2 e-book) (979-8-9925189-1-7 hardback).
1. Philosophy
2. Science 3. Spirituality 4. Metaphysics
I. Title II. Crisp, B.A. 1966- Cosmic truths book.
Non-Fiction.

Library of Congress Control Number: 2025908770

Printed in the U.S.A.

Notice: This publication is designed to provide knowledge and provoke thinking regarding quantum physics and consciousness. It includes information gathered by the author about specific aspects of technology, research, and previously classified experiments. At no time, and in no way, was classified information divulged or disseminated to the author or included in the writing of this book. Protection protocols concerning the preservation of classified materials in the interest of national security were maintained with respect to the United States of America and other countries.

Table of Contents

Part V. Ancient Knowledge & Clandestine Applications

"Out beyond ideas of wrongdoing and right doing, there is a field. I'll meet you there..."

—Rumi

Acknowledgements

Creating this work has been a freefall into a rabbit hole. There are several incredible individuals I must thank for their contributions to its final form and for yanking me back from the abyss. I appreciate the encouragement to lay my skepticism aside and suspend knee-jerk reactions to anomalies that fail to fit our established scientific or spiritual paradigms.

First and foremost, I want to express my deepest gratitude and appreciation to three outstanding researchers: Amanda J., Greg F., and Andy C.

Amanda J. for her invaluable knowledge and intelligence. Amanda has dedicated herself to exploring esoteric subjects with both analytical rigor and a healthy dose of skepticism, always striving to ground her inquiries in responsible research and scientific validation. This is no small feat, considering her extraordinary experiences—an undeniable telepathic connection with an intelligence beyond physical reality, which occurred at a neutrino detector. In a twist of synchronicity, she later discovered that the original equipment itself shared her name—Amanda!

This event set her on a path of discovery, ultimately leading her to the groundbreaking *Hypotheses* of electrical engineer Bob Beckwith, and to me. Beckwith was our shared common ground. His work suggests that humans may use neutrino light as a medium for telepathic communication, levitation, free energy and craft propulsion beyond conventional science and systems—and he was an electrical engineer. His Florida-based company was later sold to a division of Hubbel.

Amanda embraces the possibility that there is far more to reality than orthodox understanding permits. Amanda, thank you for your scientific sleuthing and your willingness to share your fascinating journey.

Likewise, I am incredibly grateful to Greg F. for his insights and contributions to my research and this work. Greg's journey began with a materialist, science-based perspective as a computer and IT expert, a ham radio operator, and as a technical equipment specialist in the television industry. A series of unusual events—first at university and later in Los Angeles—led him to explore the nature of consciousness in ways he hadn't anticipated.

His studies took him deep into the realms of precognitive dreams, psi phenomena, and the mechanics of remote viewing. As he delved further, he began connecting these experiences with older cultural references—ideas that, intriguingly, quantum mechanics may help explain.

Without formal scientific training or a prescribed framework for exploring the paranormal, Greg has approached these mysteries with an open mind and heavy skepticism. His perspective bridges the gap between scientific inquiry and the unexplored frontiers of human perception, particularly when it comes to the nature of time. Far from being the linear construct we are taught to accept; time itself remains one of the greatest and most fascinating mysteries—one that Greg continues to explore with curiosity and depth.

Greg, thank you for your thought-provoking contributions and for sharing your journey of discovery.

Acknowledgements

I am deeply grateful to Andy F. for his experiences and contributions. From an early age, Andy's life was marked by occurrences that defied everyday explanation—both in waking life and the dream world (which I might debate aren't all that divergent). Feeling different, yet searching for meaning, Andy found his first glimpses of understanding in books and classic literature during his teenage years.

His journey has been shaped by out-of-body experiences, lucid and astral dreams, and strange occurrences at home—phenomena that at times bordered on poltergeist activity. Encounters with visitors from somewhere beyond, appearing and vanishing without any apparent means of entry, only deepened his curiosity about the true nature of reality.

Through it all, Andy has been fortunate to find friends and mentors who not only understood his experiences but also helped guide him morally and intellectually—true kindred spirits to whom he feels an immense, lifelong gratitude. He credits his helpers as being like angels, shaping his path in ways that cannot be measured in material terms.

Andy, your humility speaks volumes but let there be no doubt—you are more than worthy of acknowledgment. Your adventures, perspectives, and willingness to explore the unknown have contributed something truly meaningful to this work. Thank you for sharing your journey and for making a positive impact along the way.

Again, I want to thank my editor, Michael Waitz, whose extraordinary patience and advice always help shape my books. This piece was likely his most irritating in terms of structure, arch, and length. Michael is a New Orleans native. He holds a BA in English from the University of New Orleans and an MA in Homeland Security from American Military University. Mike served ten years in US Air Force Reserve Public Affairs; wore an editor's hat for twelve years at a large legal publisher; and has enjoyed almost a decade as editor and owner of *Sticks and Stones Freelance Editing*. He and his family live near Boone, North Carolina.

Last but by no means least, my heartfelt thanks to Ghislain Viau of *Creative Publishing Book Design*. With over 30 years of experience in the publishing world, Ghislain brings unmatched expertise and genuine passion to every book he touches—from cover design to interior layout (typesetting, formatting, and eBook conversion). His insight, craftsmanship, and meticulous attention to detail gave this author the confidence to take a leap of faith, and for that, I am deeply grateful.

I acknowledge @missspookymoo for sharing her inspiration on the cover design of this book. Her vision helped me create and encapsulate the art for this far-reaching content.

I wish to thank you, the reader, for your interest and engagement with this work. It is for you that this book exists, and I hope it serves you well.

With sincere thanks,

B.A. Crisp

Introduction

The cat is dead... or it's not. It depends on who observes.

A white egret positions itself between my writing window and the sweeping vista of a tranquil lake that edges my backyard in Southwest Florida. Beyond the large pond is a thick swath of trees, where a band of coyotes' lives undercover. I hear them yipping and howling during the dark of night.

The egret launches from the grass in front of me, dipping the very tips of its wide white wings into the water. Two sets of ripples pool outward in small waves, colliding in the middle to create an eye-catching geometric pattern. They merge into a temporary swell of greater amplitude—peaks higher, troughs lower—much like my life lately: Superpositioned. Physically, I am here, but my mind inhabits other worlds, instantaneously far away and darkened, the swirling of my brain either placid or churning.

Like the wave, and these animals around me, I'm part of a greater ocean—or I am the ocean. Yet I notice just how small and inconsequential I am—the opposite of *I AM*. In noticing, I'm discomfortingly deconstructed, a fractal part of the whole shattered to pieces. Inside, I contain the entire universe—yet I'm a *have-not*. I'm also an *unknown* who has and is everything. I release a heavy sigh because trying to understand this entanglement to existence is more complicated than advanced classified calculus.

This existentialist angst compelled me to research and study quantum physics. Big, beautiful mistake. I don't recommend trying to dive below the Planck scale unless you enjoy mucky voyages into the obscure, annoying, unanswerable, and cryptic. Quantum physics is a riddle within a mysterious lattice-like matrix—a fuzzy boundary where the questions are absurd and answers elusive—unless perhaps, one has super top-secret clearances beyond *UMBRA* (code word for highly sensitive intelligence) into unknown substrates and exotic particles. I've resurfaced feeling entangled between science, philosophy, and sacred wisdom. I may require a good pair of wire cutters.

This deep dive into life's biggest questions, *"Why are we here? What is our purpose? Who are we?"* has been both a solo and a soul journey, a deeply personal and honest examination—one I took seriously until it made me laugh. The completely honest assessment eventually becomes comical—and when it turns hilarious, that's the moment Truth breaks through without fear—because fear evaporates like urine on a hot sidewalk. Our entire existence is only energy, which cannot be destroyed.

With life as we think we know it on this repository planet, we're fallible and there is no perfection. And love? It's an obvious and open secret to maneuvering the Source's (however defined) gloriously limited gift of human sentience to discover one's authenticity. As cliché as it may sound, Love heals, promotes peace, and sees best in sincere service. Thus, we are left with, not free will, as most of us would proudly or vehemently declare, but perception granted to us through some force Albert Einstein once referred to as *spooky action at a distance.*

I sit on the edge of the Everglades, where the sacred, thankfully, still holds court. And now I know: The universe speaks to me. I'm either mixed about what it says or too tethered to modern conveniences to look up and comprehend its core truths. What I do realize, sometimes, is how interconnected everything is—like egrets and coyotes—to everything else. And I'm utterly fascinated, by what lies beyond the veil of this dimension. What sets everything, all of *this*, what we think we know as *reality*, into motion?

Other questions arise: Where is my subconscious and how much of it is *me*? I highly suspect that I am not my emotions—my belongings—or my title—and I'm not my cradle-raised religion or its dogma. Nor am I what other people think about me if they consider me at all. This entire world I live in that I'm convinced I've constructed is a work of the mind. And humans are not independent parts working in isolation. But what does this *mean*?

Upon closer detached scrutiny, all the things I used to ruminate about and their accompanying emotions, pass…like gas. Life's tragedies, wants, angsts, and triumphs—each are fleeting wonders or one-off worries in a world that might look and feel real, but ultimately, is astoundingly illusory.

What I discovered is that I have no firm answers even when I'm adamant that I do. The 3-4D world performs a masterful con job, convincing most of us we have at least some of life's answers and that the ideologies and faith we adopt and promote are as solid and unbreakable as tempered glass. We insist that we *know*. We judge everything we encounter with our senses and remain completely convinced that our *perceptions* are absolute truths.

We *see* but remain blind. We *look* but fail to *observe*. We touch and go…our lives a mad dash of hustle and mayhem from birth to death. We clamor to be seen and heard—because almost nothing, we're collectively convinced, validates us more as individuals than shiny new things or influencing the masses and having them validate us in return with adoration. If only we could find that one object, gizmo, or concept that will cause us to *go viral*—and maybe feel a bit less isolated, lonely, and misunderstood.

Yet, such grasping for fame and fortune is also a fleeting and ridiculous illusion that binds us to artificial intelligence and fades in comparison to immersing ourselves in genuine creation, reflection, and flow. To most, the idea of becoming nobody is either frightening or terribly boring—and we do whatever we can to stave off irrelevance. For example, I write books that all six of my fans love.

Consequently, this flurry of industriousness and bustle we call life is less than a blip on our scale of existence. To make sense of it all we twist our pre-programmed perceptions into personalized narratives and subjective realities. We are each pleasers, conformists, rebels, and sociopaths to varying degrees. *Go!* Conquer the world and become a productive member of society, we are told. What does this mean, if society too, is a grand illusion of virtue, misinformation, obfuscation, propaganda, and suffering?

Too rarely in life do we experience bliss—snippets of extreme peace and wonder that blister our brains due to their highly concentrated, unforgettable essence. We could use more bliss and peace. Most of us instead, cling to our suffering and judgments, finding it an addictive alternative against elevated transcendence.

As I float and trample my way through this quantum adventure, it suddenly occurs to me that life is likely a complete fiction… as an author, I conjure. I imagine characters and create entire universes in my head by mentally constructing plots and settings—a proverbial house of cards if I screw up arcs or include insincere passages. Fiction must encompass relatable human drama combined with ulterior thinking for it to be worth reading.

As sentient beings striving to keep our biological containers intact on this planet for as long as possible—human beings do the same thing—we strive to master fiction. Our lives are a fiction we sell ourselves and others, so we may adapt our perceptions to the interface of this reality, which doesn't really exist—except in our minds (which often fools us), brought to us by consciousness that is kicked off by God…only knows what, beneath a Planck Scale.

What is so utterly fantastic is that because our lives are fictitious illusions based on our perceptions, the universe offers us a plasmic platter from which we may create our worlds through as many redrafts as we might imagine. More on that later.

My foray into quantum physics alongside a crash course into subatomic worlds has forced me to tread some weird, discomforting, and mysterious aspects of *being*. Am I full of shit and delusional? Or am I a genius? Do we really exist? Or are we part of an artificial but intelligently conscious simulation? And—I'm not so convinced there was a Big Bang or that dark energy exists. What I do know is that humans are, as a species, one undulating and observant quantum field, impermanent, dynamic, and complex.

According to Hugh Everett III's *many-worlds interpretation*, all possible outcomes may be true, albeit in separate but parallel universes. Again, *perception*. Legions of human beings occupying this planet equals countless perceptions. These perceptions are brought to us by unseen waves of zeal that collapse into *thoughts*, which gives rise to *energy, feelings,* and sometimes, *action*.

Humans make and break universal laws based on perceptions. We adjust our behaviors, and our feelings based on perceptions—whether accurate or not. We judge others and assign sins, sacredness, or science to them based on perceptions (and programming).

Perception is our captor and our liberator. In contemplating life's mysteries, it spurs creation, destruction, suffering, liberation, inaction, or idleness. And it keeps us firmly locked in the cosmological gravity of almost laughable proportions.

Consequently, Alexander Wendt's *Social Theory of International Politics* emphasizes our foolishness. He claims it is our shared ideas and norms, which shape behaviors of state, albeit, helped along by propaganda and misinformation—a form of skittish collective consciousness moving massive amounts of subconscious information and energy.

Wendt believed that state identities and interests are constructed by social structures, not by human nature or domestic politics. He called the views we hold "cultures of anarchy". Basically, society and culture shapes us...for better or for worse—but mostly for worse—right down to our quantum core. Global collapse happens through us and finalizes reality based on our perceptions—the ultimate immeasurable thing. Conversely, universal cohesion emerges beyond us and initiates reality independent of our perceptions—the ultimate quantifiable constant.

Depending on my own or someone else's *perception*, I'm a charlatan, an angel, a looney, a genius, a scribe, a shaman or delusional. I'm likable and I'm loathed. I'm none and all these things at once because humans typically *infer* and collapse on a judgment based on personalized probabilities, rumors, or other forms of societal coding. I fall into a label and remain there while at the same time existing someplace else under another label, a mirror to the positive or negative charge—a single photon of nonlocal potential entangled with trillions of others.

So, if a planet full of people are going to naturally subscribe to their perceptions regardless, especially the negative ones, since we are impelled by our biology and nature to do so, why not take some risks and ask life's probing questions? If we are created in the image of God (or Gods as the Christian Bible so confusingly states in Genesis using the term, '*our* image'), we are made to be curious, creative, and to explore. So, here it goes:

How can a particle be everywhere and nowhere at once? Does a thought travel faster than the speed of light? Is there life after death? Is telepathy possible? Is Nikola Tesla's particle beam weaponry now part of a secret military arsenal? Are neutrinos the key to consciousness and instantaneous interstellar travel? Does prayer work? Does this planet we occupy reside inside a black hole? Is space-time a fiction too? Are we so dominated by gravitational pushes and pulls that we've overlooked or ignored the electrical universe in which we're immersed? Do I have enough toilet paper?

Alternatively, vivid dreams of strange spaces awaken me at night. Some sweep me into a world of light waves, vibrational energy, and geometric patterns, while other dreams revolve around me trying to make a flight, interpret my uncle's mathematical equations, tame wildcats, or being unable to pass a shaman's exam. But the most vivid dream I have is passing through a portal into a star system, where two planets, and two suns, orbiting in proximity to one another, host life... and I feel as if I've come home.

While awake, I'm suspended between two worlds—one spiritually abstract and one quite physical—where a stubbed toe and my Zen empty "Fuck!" feels painfully human, solid, and real. When I meditate, the chatter of this worldly illusion and its suffering becomes more like steam and disappears. Then, I'm not *here* even though anyone who looks in on me would vehemently counter that I'm sitting right *there*.

Moreover, I occasionally ask myself: Am I dying? Is this feeling of suspension between two worlds, like straddling a supernatural fence, the beginning of my end? The answer, I remind myself, is that we're all dying. Again, and again. It's a fact that most people in this world abhor—we die. But what is death? Some call it 'The End'. Is it? Maybe death is really the *beginning*.

Mystery schools and secret societies view death as a good thing and subject initiates to 'crossing over', a form of collapse or transition of the old self into a state of rebirth, where typically, on the third day, the initiate rises again after being pulled out of a ceremonial snake pit or makeshift coffin (true story...stay tuned).

Christians say that when someone dies, they've *gone home* to be with Jesus. This should be considered a 'Happily Ever-After' event but instead, the *born-again* mourn for themselves if not the event. We turn away from death as if it is a curse. Perhaps it's more of a curse to live if one cannot or does not know how to *perceive* well. And if a part of us continues after death, why not celebrate such a glorious transition? I assume it's because death has little to do with the person who's moved on and plenty to do with our hearts, a vast vault of quantum memory and emotion—we grieve the abrupt and eternal physical loss of someone we care about—a human or pet we will always love. While the grave might separate our energies it is Love that keeps us eternally and quantumly connected.

Buddhists and quantum physics tell us there is no death—only transition of energy...but to what? And where? It's challenging for the human ego to ponder the complete cessation of its illusory independent and personalized existence. When I die, I still want to be *me*. Yet I have no idea what this means because I'm still, after decades on this planet, learning about who *I am*.

I realize these questions are headlong leaps into a head spinning vortex, and they were mostly prohibited during my youth. Organized religion strongly encourages strict obedience and blind faith. It fervently disfavors questions or contrary evidence among the masses (pun intended) …but so does science.

For example, if the outcome of a scientific experiment fundamentally conflicts with 'well-established physical laws' (the very phrase holding a level of inherent bias), it is vehemently discarded—or pushed into the shadows. We seem to be more comfortable laying anomalies aside rather than examining them. This much, science and religion have in common. Like death, we turn away from anomalies, especially if they threaten paradigms we've been programmed to accept without wonder.

Shamans might call my foray into this largely secretive arena a "first-eye" and "first-ear" quantum quest. Priests might accuse me of occult practices, even though they privately engage cryptic rites much more than I. Politicians could use the book as fodder in any way that might further their careers or secretly fatten their wallets. Scientists might hypothesize that I'm an unfortunate victim of procedural entropy. And who knows? Perhaps the military-industrial-intelligence complex will add me to their secret list of people who bear watching.

But no one need worry.

I harbor zero evil plans (or delusions) to topple any governments so I may rule the masses in the throes of some grandiose cosmological revolution. Rest assured, it has taken me a lifetime to master becoming nobody—to stand back and let others push, elbow, and shove their way to inevitable death or practice false humility. It's a fact that no matter how much we negotiate against our end or work to stave off the grim reaper, it arrives for us all—and just as we came into this world alone, so too, we leave it the same way.

It is unknown for certain among the living if we dive headfirst into some great abyss at death, or if we are released from the matrix into a Godly paradise of cosmological proportions, where answers to life's most unanswerable questions are bestowed upon us like gold in perfectly manicured gardens by previously departed loved ones.

"You're not qualified to write about science or quantum physics!" I was told. "Let this book die."

Is anyone qualified to write about quantum physics?

The good news (or maybe it's bad news) is that scientific and supernatural paradigms, outliers, and anomalies, like fractals and mostly unknown penniless authors, can and often, are shattered. Ignoring or scoffing at them doesn't make them less valid or part of the whole. And ego, like entanglement, I've learned, can be two places at once: arrogantly overconfident, territorial, and falsely all-knowing, or absurdly doubtful and insecure.

This book is a neophyte's attempt to create a more focused gateway between science and the sacred. Here, I bring together a great amount of phenomenal quantum information and compile it within a manageable portal of interest. I follow the foundations of quantum physics into the abstract, through the study of energy, plasma, mathematics, information theory, human potential, and integration. I dive into DNA, declassified intelligence and military files, faith, spiritualism, the human brain, neutrinos, God, and consciousness, among a host of other topics—and I've interviewed people with clearances so high (or low), that like quantum physics, they might be beyond our conscious horizon.

Consider this work a manifesto of mystery, an encyclopedia of the enigmatic, a journal of scientific odyssey, a grimoire of glitches, and the unholy grail for a guild of new far-thinking grandmasters and guardians. Contribute to it, take from it, or leave it, according to your *perception* and consider it, first and foremost, a work of **fiction**.

This simple book about convoluted subjects is designed to be read from any chapter and is, by no means, comprehensive. I know I will have missed a lot of important research, ideas, and people. Consider this book, *A Cosmic Conspiracy* your key to expansive fields of the known and unknown, open for debate, interpretation, and for further editions, but for God's sake, not gospel.

In these pages, a cat will be skinned. But rest assured, no animals were harmed during the creation of this book.

PART I

Foundations of "Reality"

CHAPTER ONE

The Dance of Two Worlds:
An Unofficial and Rudimentary Primer on Classical & Quantum Mechanics

L eonardo da Vinci never attended university. He had no formal education, which he considered a blessing. Da Vinci was not degreed in any field. Driven by curiosity and a powerful imagination, he considered himself a "disciple of experience". He used notebooks, some 40,000 of them, to record his observances of nature and his ideas.

Da Vinci *anticipated* future scientific discoveries and inventions. He'd sketch them out in detail. Today, he is still credited with being an expert artist, musician, engineer, inventor, architect, and scientist.

I am not da Vinci. Read into that statement what you will. While I have a formal education and a post graduate degree, I'm not a scientist, an engineer, a musician, an architect or an inventor...unless you count me an inventor and architect of stories. I am a writer of fiction. And I have boobs.

Physics, life, and fiction appear, at least to this observer, to have much in common. And like da Vinci, I am a curious observer who reads constantly and records all manner of facts, trivia, ruminations, drawings, and feelings into notebooks or computer drives. I also study people and once worked briefly with the *criminally insane* as a psychiatric assessment specialist. My brain is *occasionally* tapped by governmental agencies and private contractors for strategic intelligence sessions. This basically means I know how to shovel all sorts of shit but am mostly disarmingly honest.

Similarly, DNA tests show I have multiple Viking warriors for ancestors, come from a long line of Italian and German craftsman and engineers, and am possibly a remnant of Ottoman Empire lineage due to deeply buried Turkish roots. Yet these descriptions do not define me. They literally mean and prove nothing at all to anyone… unless…I or you, *perceive* them to mean something.

Plus, I sometimes hear my grandmother whispering from the grave, "Don't let the weight of your head knock you on your ass!" This haunting voice is more likely to keep me programmed in the linear line of reason and established paradigms than any other cautionary warning—although I've learned to turn around and hug her ghost—and keep asking questions.

For some odd and nearly obsessive reason, quantum physics, especially where it becomes entangled with the sacred, consumes me. When I pluck a blade of grass from the yard, where exactly, does the blade of grass begin and end? Where is the sun in this blade of grass? Where is the water? What makes it a blade of grass instead of, say, a fingernail? Each is unique until we tunnel more deeply into the quantum makeup of matter, where God, ghosts, and gallbladders have everything in common.

The same might be asked for any sentient or inanimate thing on this planet. Sentient beings and inanimate objects on Earth and other planets are, I'm convinced, conduits between macro and microscales of existence. Living things and inanimate objects are the fractal bridges of energy between the seen and unseen, driving consciousness through everything we touch, manipulate, or think about.

In the everyday world, most humans are sure that matter is solid, something we can see, feel, taste, touch, create or destroy. Anything composed of physical substance readily available to our manipulation and baseline senses falls under the purview of matter. Matter is something, according to scientists, that has mass and takes up space. Some examples of this are solids (like blades of grass), liquids, plasma, and certain exotic condensates (Fermionic, Exciton, or Bose-Einstein).

I blame Rene Descartes and his Cartesian Model for this limiting perspective, which has served as both a blessing and a curse for this earthly illusion passing as a stubborn nondebatable reality. Descartes concluded that body and mind, physicality and spirituality, belong to separate realms. Simply put, if something cannot be repeatedly tested, held, weighed, measured, and replicated, it is bullshit.

I completely understand why Descartes adopted this *dualism*. Back in the early 1600's his contemporaries were being beheaded, hanged, or burned at the stake if they proposed anything scientific or otherwise that might trespass against religion and its leaders. A deal was struck. Science would adhere to the material world and what could be *seen,* while religion would have domain over everything *unseen.* For centuries, this has worked out—especially for the Church leaders and the scientists.

Each was free to explore the seen and unseen according to subjective interpretations or experimental results while keeping the masses in the dark—at least until Nikola Tesla came along! This separation of science from the sacred is likely the longest-running covert operation the world has never known. And it's been successfully sold as a bill of goods that props up capitalism, *and* the promise of a heavenly eternity, while largely omitting spirituality in favor of dogma, ongoing wars, and fleeting materialism.

Consequently, the most beneficial scientific and sacred discoveries remain secret until they become obsolete and something more advanced is discovered—and the former only then get packed up like shiny new toys, rebranded and distributed among the lowly masses. So, according to what we think we know, the most fundamental matter is composed of atoms, which are, allegedly, made up of protons, neutrons and electrons. Each may be measured—with a caveat.

Humans not privy to classified mathematics cannot simultaneously measure position and momentum of atoms or parts of atoms with perfect precision. The very act of measuring affects the particle causing it to change, as if its aware and *thinking*. Furthermore, and as far as we know, the smallest bit of matter in our universe is the neutrino (unless we count hypothetical Axions). But we can't directly measure a neutrino due to its negligible mass and neutral charge. Or can we?

In a physical world, and according to acceptable, established science, we cannot measure anything smaller than a Planck length without literally producing a black hole. But what if we already reside in a black hole? Black holes, I'm learning, exist anywhere, like the inside of a rock, and come in a variety of sizes and intellect—if they truly exist. Or maybe we made a dispensation for black holes within the gravity model of physics so other things would "fit" and purposely drive us away from plasma cosmology.

The problem is that within the realm of quantum gravity, behavior at such small scales can't be observed. Yet beyond our ability to measure, we can *infer* or vertically *transcend*, which swiftly takes us out of the realm of science—or science as most humans without high clearances think they know and understand *science.*

Albert Einstein taught us that matter and energy are interchangeable through his gloriously concise equation $E = mc^2$. This means that *any* type of matter can take a different form. So, does $mc^2 = E$? Not exactly until we do this: $m = E/c^2$. I'm told it's the same thing. Not quite. In a physical 3-4D world such as ours, his equation works well—until it doesn't.

The very same equation, which insists matter and energy are interchangeable also renders the traditional definition of matter much more complex. For example, when matter is converted to energy (nuclear fission) or vice-versa (particle accelerators) it severely blurs the lines of what constitutes matter and energy.

Matter is more than "stuff" and energy creates matter. For instance, a photon lacks rest mass but has energy—so, is it matter? A proton gets mass from the energy bound inside it—but not from its quarks— so, is it energy? Or does it behave more like the mass and energy of a brain microtubule?

It is also important to remember that matter is comprised of subatomic particles; namely neutrons, protons and electrons inside an atom—which is made up of—oscillating energy grids. Yet, electrons can be found independent of atoms. What now? It might be easier to say that particles are states of a field where potentials actualize.

Quite interestingly, from this vantage point, what appears to us as matter is really nothing more than energy vibrating at different rates of oscillation. And oscillation is very important because vortexes matter. If everything goes static, or remains perfectly still, life eventually ceases. On the other hand, oscillating energy opens multiple possibilities and worlds most human beings have never imagined.

Consequently, dark matter is another matter. Scientists know it exists due to its gravitational effect on visible matter, but humans have never been able to find it. And despite my admiration for Einstein and Newton, I can't help but blame each of them for imprisoning us in gravitational theory.

Newton supplied instantaneous gravity to Einstein's strict speed limit of matter, which seem to counter one another. Each skirt the issue of why inertial mass is equal to gravitational mass, and gravity doesn't appear to involve time. This does not *feel* right.

Furthermore, Einstein introduced sci-fi lovers everywhere to the idea of warped space. Neither he or Newton could reconcile gravity with electromagnetism or the electrical nature of matter and the universe...but I suspect that gravity is much more amenable to mathematical models than electricity—and it is the latter which gives rise to gravity. And gravity too, is its own conundrum. We do not even know exactly how gravity works or from where it arrives. I suspect it is a byproduct of electromagnetism, yet I have no mathematical equations to support such a guess. Somehow, I sense we've been steered away from our electro-magnetic connection to Earth and the universe and rerouted by forces far beyond 'need-to-know' so we firmly remain under the weight of gravity.

I must give Einstein credit for publicly admitting his theories were not ironclad but more like sheets to be thrown over ghosts—except the sheets come in a variety of clandestine agency insignias and black holes can be cut into them almost anywhere.

An example of dark matter's existence is that the outer regions of galaxies rotate much faster than they should. Where dark matter becomes both murky and quirky resides in what we *think* we observe:

- Gravitational lensing: Light from distant galaxies bends more than it should around large objects, suggesting unseen mass
- Cosmic microwave background radiation patterns
- The structure and formation of galaxy clusters
- The overall large-scale structure of the universe

Dark matter are allegedly particles that make up approximately twenty-seven percent of the universe (dark energy is considered a *force,* which makes up sixty-eight to seventy percent of the universe, causing it to expand, and is not considered the same as dark matter). Dark matter does not emit, absorb, or reflect light or other electromagnetic radiation and interacts with ordinary matter only through gravity, as it appears to be distributed in large halos around galaxies.

Composition theories of dark matter

These include yet-undiscovered particles such as:
- WIMPs (Weakly Interacting Massive Particles)
- Axions (hypothetical very light particles)
- Sterile neutrinos

Despite numerous search and rescue attempts, humans have never directly detected or observed dark matter (or dark energy) particles—they're barely able to capture neutrinos (we're told). Alternatively, many scientists are conducting research, which may prove that dark matter doesn't exist, or if it does exist, that its more stable than we thought—Hubble tension and fudge factors be damned.

We are missing *something* in the cosmological model. The Universe, we've learned, remains complicated. This is what *we,* meaning 'Joe and Jane Public', have been told by some of the greatest science minds in the world. But those great science minds can't be totally blamed—they were taught a certain way and ordered to restrain their research and experiments only to that which can be measured—unless they become part of a secret society or mystery school.

Consequently, scientists use various detection methods to locate dark matter including:

- Underground detectors seeking rare collisions with normal matter
- Particle accelerators hoping to create dark matter
- Space-based telescopes studying dark matter's gravitational effects

The nature of dark matter remains elusive in physics, and its true discovery would represent a most extraordinary breakthrough in human understanding of the universe—so much so, that it would likely upend every aspect of life that the human species accepts as scientific or spiritual gospel. This is where other scientists, those still under strict security oaths aren't talking...because they can't.

The time-scape theory, another recent alternative hypothesis, suggests that *dark energy*, could be an illusion arising from our mathematical treatment of spacetime measurements and cosmic expansion calculations. Large voids and dense regions of the universe might account for our *perception* of an ever-expanding universe. In other words, according to scientists from the University of Canterbury in New Zealand, our universe is lumpy and kinetic-energy dependent—where time might flow slower in dense regions and faster in voids.

But that's not all...in March 2025 an international team of astronomers running the Dark Energy Spectroscopic Instrument, or DESI, at Kitt Peak National Observatory in Arizona, discovered that dark energy is not a constant force of nature but one that ebbs and flows through cosmic time. Simply put, this means our universe will likely be spared from being torn apart as its expansion wanes—or maybe, reverses course.

Interestingly, a separate result from the Atacama Cosmology Telescope in Chile, reinforced DESI findings. This has led to more than a bit of cosmic confusion over what this means for Hubble tension (the discrepancy between different measurements of the Hubble constant (H_0) — the rate at which the universe is expanding.

But wait.

I began this chapter trying to explain *matter*. How did we suddenly make the invisible leap from the physical into something less so? What is the spiritual, not necessarily in the *angels sent me here* sense, but in the quantum realm?

Comfortable or not, everything we think we know as solid, even if it's a chair, a person, science, or religion, is governed by the quantum, having a resonance of universal spirit. And the smaller one quantumly tunnels, the greater energy, spirit, and confusion we find.

Now, let's return to the 3-D world for a moment to get a grip on something more solid:

Classical Physics (A simpleton's version)

Classical physics is a broad term that encompasses about four major areas of physics developed mostly during the last century. At its most basic, classical physics includes:

1. ***Classical mechanics*** - The study of motion, forces, and energy of macroscopic (larger) objects, predominantly brought to our attention by Sir Isaac Newton. It focuses specifically on describing the motion of objects under various forces. It incorporates concepts such as: a) Newton's laws of motion, b)

conservation laws, c) gravitational theory, and d) the mechanics of rigid bodies and fluids.

2. ***Classical electromagnetism*** - The study of electricity, magnetism, and light as described by James Clarke Maxwell's stubbornly entrenched mathematical equations, which launched the human species into what remains (outside of the military-industrial-intelligence complex) a rudimentary technical age.

3. ***Classical thermodynamics*** - The study of heat, energy, and the behavior of matter in bulk

4. ***Classical optics*** - The study of light behavior through reflection, refraction, diffraction, etc.

Both classical mechanics and classical physics work extremely well for everyday 3-4D earth-based situations, and this is why experts in engineering and regular everyday people adhere to them. However, each break down when dealing with:

- Very high speeds (where special relativity becomes important)
- Very small scales (where quantum mechanics is needed)
- Very strong gravitational fields (where general relativity is required)

Despite this small hiccup, it is important to note that electrical engineers have worked in orthogonal six-dimensional spaces, meaning that all dimensions go from negative infinity to positive infinity with right-hand relations between the two—or—voltage gradients and magnetic fields.

So, we have depth, width, length, space-time, magnetic fields and voltages, the latter being the "pressure" that *pushes* electric current through a circuit. Yet we largely ignore voltages because gravity models of the universe do best in an electrically sterile universe. And opposition to electrical theories of the universe exists at plenum density. If it didn't, and we entertained an electrical universe, it would bankrupt some people who consider themselves much more important than the rest of us—and those highly important people do not want to be 'poor folk' who suddenly find themselves irrelevant.

Quantum Physics aka Quantum Mechanics (Matter and Energy at its Smallest Scales)

Quantum physics (or quantum mechanics) is a fundamental theory in physics that describes the behavior of matter and energy at the smallest scales—typically at the atomic and subatomic level. It's radically different from classical physics in extraordinarily supernatural ways:

Wave-Particle Duality: Everything (particles, light, etc.) can behave as both a wave and a particle, depending upon how we *observe* it. For example, electrons can create interference patterns like waves but can also behave as discrete particles. They're everywhere at once until we look at them, and observation leads us to:

Superposition. Objects can exist in multiple states simultaneously until they're measured or observed. This is famously illustrated by Schrödinger's cat, a 'thought experiment'. Superposition refers to a quantum system existing in multiple states simultaneously until measured.

A classic example is the electron spin that can be in a superposition of "up" and "down" states at the same time. Consider it a game of *heads or tails* void of probability or chance. For example, an electron's spin really is both up *AND* down at the same time, until measured. It's **not** a clone or two separate pieces that suddenly bind together. It is the same thing existing in many places at once.

This is not uncertainty or incomplete knowledge but a fundamental feature of quantum reality—and it is frustrating for anyone *asleep in the dream* to understand. Particles can exist in multiple places at once and shapeshift or become entangled.

Yet while mathematical rules (*linear combination, born rule, unitary evolution, and normalization condition*) govern how superpositions evolve, they don't fully explain the measurement problem—why and how superpositions collapse into definite states upon measurement.

Schroedinger's Cat

Erwin Schrödinger was a Nobel Prize–winning Austrian physicist who developed fundamental results in quantum theory. But Max Born, like Nikola Tesla, is often left out of this equation. While Schrödinger gave us the cat, it was Born who gave us his interpretation of the wave function (ψ) in Schrödinger's equation.

This idea—that the outcomes in quantum mechanics are fundamentally **probabilistic** rather than deterministic—revolutionized physics. Before Born, it was assumed particles had definite trajectories. After Born, quantum physics became a theory of **uncertainty** based in individual particle probabilities.

In Schrödinger's 1935 thought experiment, a cat is sealed in a box, along with a flask of radioactive poison and a radioactive monitor. While the cat remains sealed in the box, it is theoretically both alive and dead. (The same cat we're trying to skin in this book…or not.)

The cat's fate is linked to a subatomic event that may or may not occur. This is known as *randomness* because any outcome is possible. Randomness basically means there is no order or pattern. Quantum mechanics dictates that randomness is a fundamental aspect of nature. If randomness were lacking in nature, diversity would plummet. Humans and other biodiverse life would lose ranges of possibility and eventually become far less adaptable to homogenous environments making them susceptible to disease. DNA would have limited options for natural selection, eventually leading to extinction.

So, fundamentally, this experiment asks when and how long a subatomic particle/s remains in multiple places at once before it collapses into a dead (or not) cat. In other words, where does super-position end and our understanding of *reality* arrive to give us only one outcome? Once *observed*, the quantum possibilities *collapse* into a single definite state. Simply put, humans influence particles, and therefore, events, energy, and matter.

For a sci-fi writer like me, this begs the question: What, for fuck's sake, is real and do we have free will? It also enables me to lie my ass off while writing sci-fi—provided I make it believable. When composing fiction, I'm able to take great liberties with science, so long as it remains within the realm of fiction, using *imagination*. I'd initially thought I had an *aha!* moment regarding quantum mechanics by putting superposition on equal footing with potentiality.

"Be very careful," I was told, because superposition has specific mathematical and physical meanings that don't perfectly map onto the philosophical concept of potentiality. I'm not so sure I agree. Something *feels* off about that. But *feelings* are considered far outside the scope of a testable hypothesis and fall on the dull plastic sword of empiricism.

David Bohm, a physics philosopher, has explored these very connections between potentiality, superposition, and the deeper links between quantum physics and philosophy, which reminded me again, that there is no new thing under the sun, despite any egoistic human tendencies to perceive my ideas as completely original and Nobel worthy.

Bohm's most profound contributions to philosophy (and I would argue physics), attempt to bridge at least one gap between the quantum and the classical. Bohm proposed that **The Implicate Order** is a deeper level of reality where everything is interconnected and continuously in flux. The **Explicate Order** is our everyday world of separate objects and events. **Reality,** according to Boehm, constantly unfolds from implicate to explicate and enfolds back again. In other words, the universe is undivided in one totality, something he called the *holomovement.*

Consequently, for Bohm, space is not an empty vacuum but a plenum—full and teeming—leading him to postulate, quite elegantly through mathematics, the existence of a vast hidden realm. His findings suggest that randomness fades when context becomes sufficiently deep or wide, revealing that what appears as randomness is incidental rather than fundamental.

As we'll explore in Chapter Four, Bohm's ideas—particularly his theories of hidden variables and the implicate/explicate order—were largely overlooked and underappreciated by mainstream science for decades. This neglect stemmed, in part, from the dominance of empiricism, which prioritizes strict control and predictability over the pursuit of universal truths—truths that concepts like holomovement and wholeness, inherently unobservable in a laboratory setting, attempt to express. But there's more.

Likewise, scientists cannot escape the fact that Bohm's hypothesis was and is rigorously grounded in much later and advanced experimental physics, resulting not in wishful thinking but an entirely new physics—one that encompasses the science with the sacred.

Let's meet another noteworthy gentleman who contributed to our complete misunderstanding of quantum physics with a wildly intelligent theory. Developed by Werner Karl Heisenberg, a Nobel prize winning pioneer in quantum mechanics, the ***Uncertainty Principle*** states that we cannot simultaneously know both the exact position and momentum of a particle with perfect precision. The more precisely we measure one, the less precisely we know the other. This holds true up to a tipping point—one we'll arrive at later in this book.

This leads to some squirrelly, head-scratching moments, which spiral us right into ***Quantum Entanglement***, where particles are instantaneously non-locally connected. This means that measuring one subatomic particle instantly affects the state of the other, regardless of distance. Einstein called this "spooky action at a distance."

Entanglement makes me wonder, like Bohm, if the particles are not as separate as nature and the militarized-intelligence-corporate complex leads us to perceive but is actually only one particle—yet in wondering such an outrageous thing, it might mean that EVERYTHING we think of as separate, and as matter, divided by space—is *ONE* thing. If this is true, entanglement has some seriously humorous implications for the idea that duality is a necessary part of human existence in our 3-4D world.

For example, if you and a friend were entangled, you could get drunk without drinking a drop of alcohol. Every time your friend does a shot, you'd instantly "know and feel" it. Downside? If they throw up and pass out, so do you. But what if this entangled energy is less direct and more subtle, such as a wave of thought, information, or emotion moving through a crowd or a series of computers, influencing and directing this holographic illusion we think of as reality? I digress. Emotion likely doesn't move through computers the way information does—it can be translated but not replicated in the same manner as information.

Quantum Tunneling happens when particles pass through barriers that classical physics says they shouldn't, like the enzyme glucose oxidase, which transfers electrons during the breakdown of glucose in living cells. Without quantum tunneling, biological processes would prove impossibly slow, if not come to a complete halt.

Alpha radiation is another example of quantum tunneling where an electromagnetic wall inside an atomic nucleus should prevent alpha particles (two protons and two neutrons) from crossing the barrier—but it doesn't.

Quantum tunneling, like other quantum mechanical properties, is largely a baffling mystery—the stuff of sages and sorcerers—or extraordinarily advanced thinkers in clandestine agencies who might have seriously beefed up their thalamocortical systems. (The thalamocortical system consists of the thalamus, a structure deep within the brain, and the cerebral cortex, the outer layer responsible for higher-level cognitive functions. Some covert projects have included understanding and modulating its activity, leading to greatly enhanced intelligence and cognitive performance).

This analogy of quantum tunneling may help: In classical physics, if we roll a ball uphill and it loses acceleration before reaching the top, we naturally expect it to roll back toward us. During quantum tunneling, the ball could possibly appear on the far side of the hill while also rolling back to us!

Thus far, QT only works at the subatomic level and cannot be scaled up to beach balls or UFO's. Or can they? Notably, *Quantum Tunneling* is not to be confused with wormholes, which we'll examine later in this book.

Quantization is where certain physical properties come in discrete units rather than varying continuously. Energy, for instance, is exchanged in discrete packets called *quanta,* the smallest indivisible unit of a physical quantity in a system.

Electromagnetic radiation is also known as *light* by non-scientist types. Its smallest indivisible unit is called a photon. However, in quantum mechanics (QM), particles like photons (and others) can bind their fates together, even at opposite ends of a galaxy, which means that *they* may not be a '*they*' at all but an *"it"*.

Astoundingly, there may be no true separation of particles, and all particles in existence, including humans, might be one particle—or not even a particle at all. We might be part of a void that only gives off an impression of movement because we're trapped within it. This lends credence and perhaps coherence to the unnerving idea that everything is interconnected, and we, and everything we think of as *real* or *separate* is a holographic fragment of our imagination—or a reflective wave of potentiality mirroring someone (or something) else's reality. We don't truly exist in the sense that we think we do.

I *perceive* that Bohm was onto something.

If this comes as a major shock to the 3-4D limited ego, just imagine the upset scientists and the self-sanctified feel!

Alternatively, sages, saints, and seers are less bothered by such mysterious voids and undulating waves, having long ago realized the nature of our *reality* through multiple trips into transcendent states of awareness. Each has discovered that we are surrounded by multiple dimensions—and there exist *beings* in those dimensions. Don't believe me? Where then, do your angels and God reside? If you tell me they're in Heaven, where be this Heaven, exactly? If you tell me God doesn't exist—you're in for a real 'first-eye'-opener as you read on!

No scientist worth his or her weight in tenure or secret clearances dares to freely explore where shamans, monks, philosophers, seekers, priests, believers, sci-fi writers or remote viewers (aka *extension neuro-sensors*) freely tread. This sort of *spooky action at a distance,* used loosely here, drives scientists' bat-shit crazy because in a strictly mechanical, material, and measurable world, quantum mechanics makes zero sense, which will lead us to Zero-point energy…but not yet.

The Anthropic Principle (AP)

Brandon Carter proposed the AP in 1973 and claims that the universe we live in is finely tuned to specifically host life. AP can be formulated in different ways like a *description* akin to "I think, therefore, I am" to a *knowing* such as claiming that the universe evolved for our existence. The latter claim is considered more philosophical than scientific, and the AP is considered a poor scientific tool because it can't be measured.

Briefly, three scientists stepped forward in 2024 with a way to measure the AP by combining three elements: cosmic inflation, dark matter, and axions—and measuring each using something called the LiteBIRD satellite for the study of B-mode Polarization by the Japanese Aerospace Exploration Agency (JAXA). It's expected date of launch is 2032. In essence, hypothetical axions could be dark matter—or not—leading to **a)** a failure of AP where different rules govern the universe, **b)** a governance under new dynamics not yet known or understood or **c)** the true nature of cosmology is more complicated than we realize.

But what if the true nature of cosmology is simpler than we think? What if cosmological answers are staring us in the face like a ketchup bottle in the refrigerator, one we fail to see right in front of us? Perhaps all it takes is an outsider, one untainted by established paradigms, to unlock and free the quantum field in the bottleneck.

QM principles (AP being the exception) exhibit behaviors that defy our everyday logic but have been repeatedly confirmed through scientific experiments. Furthermore, quantum physics has practical applications in:

- Lasers
- Transistors and modern electronics
- Quantum computers
- Medical imaging (MRI)
- Nuclear power
- Solar cells

Now we come to the fascinating, yet even more irritating part of quantum mechanics known as the *Observer Effect*. The observer effect describes how the mere act of measuring or observing a quantum system inevitably affects and changes that system. It's like asking if a tree falls over in a forest, does it make any sound if a sentient observer isn't there to hear it? Yet the observer effect is fundamentally different from classical physics, where observation doesn't inherently disturb what's being measured. Here's why this happens:

- To observe something at the quantum level, we must interact with it in some physical way, using a probe, our eyes, or other detection method
- Even something as gentle as bouncing a single photon off a particle to "see" it will change the particle's state
- Contrary to some social media influencers, this isn't about human consciousness or observation—any interaction, i.e. a camera, a person, or a mirror, that could reveal information about the system, counts as an "observation"

- The mystery isn't that observation changes things (which happens in classical physics too), but rather why and how quantum superpositions collapse into definite states upon observation. *How does it know?* This leads many to believe or infer that we live in a consciously intelligent universe, one of energy, which moves outside of time or space.

The Annoying Double-Slit *Experiment*

When electrons are fired at two slits without measuring which slit they go through, they create an interference pattern suggesting they went through both slits simultaneously (acting like waves). Yet when we try to measure which slit each electron goes through, the interference pattern disappears, and the electrons behave like particles. Consequently, the act of measurement forces the electron to "choose" a definite path.

Conversely, the *observer effect* is often confused with the *uncertainty principle*, but they're different because the observer effect is about how measurement disturbs a system, while the uncertainty principle is about fundamental limits to what can be known about a system, regardless of how carefully we measure.

The double-slit experiment highlights the strange, counterintuitive nature of the quantum world where the line between observer and the observed seems to purposely or knowingly blur. But how and why?

To complicate the already convoluted world of quantum physics as humans understand it (or don't), its system appears to start as a wave function that represents all possible states at once—or what is known going forth in this book as *Bohmian potentiality*:

- These states coexist until a physical interaction, i.e. a measurement of observation, forces the system to "choose" one state

- The interaction of an observer/measurement creates quantum decoherence where the system becomes *entangled* with the measuring apparatus
- This entanglement effectively forces the superposition to collapse into a classical, definite state

In the double-slit experiment, without measurement, electrons behave like waves, going through both slits—but with measurement, the interaction at one slit force the electron to behave like a particle. The very act of measuring requires an interaction that destroys the wave-like behavior (it collapses).

According to physicists, quantum mechanics is indeed a freefall into the supernatural abyss. The best minds in the world can't connect or understand how or why particles entangle themselves, superposition, or act as a wave. To even ask the questions seem nonsensical, like trying to put together a super-difficult multi-billion-piece monochromatic puzzle of a void.

However, I did locate an interesting website operated by particle physicist Matt Strassler called, *Of Particular Significance,* where he does an excellent job of explaining why elementary particles *do not* exist.

Likewise, all of us who dare consider such heady topics are on a quest to answer ancient questions and to unify the subatomic world with our material world—to find the switch that lights up all levels and dimensions within our universe—so that we may both see and understand what or who made us—or if we are even real!

CHAPTER TWO

Schrödinger's Library Card
How to Access Information in Every Universe
(Without Late Fees in Any Dimension)

Quantum physics is a gorgeously exciting exercise in futility where humans often extract ideas, inventions, and elevated philosophies—but little else. If one expects to find concrete answers in a vat of goop, insanity and frustration are its consolation prizes.

It appears that subatomic particles and wave forms are the gatekeepers of a mysterious substrate—the spooky stuff that lies somewhere beyond human comprehension and ignites life. Attempting to understand quantum physics is like meandering through a labyrinth of dead ends, riddles, and ridiculous questions—and then trying to saddle phantoms and photons.

Before we jump into this *"splooge"* to examine how multiple universes might simultaneously exist across eleven (or more) dimensions while spontaneously collapsing into reality on infinitely large cosmic membranes, we should probably consider a lobotomy in our own dimension (and possibly several more in parallel ones). Never one to back down from a challenge, let's dive into the muck.

The *Many Worlds interpretation*, *M Theory*, *Brane Cosmology*, *GRW Theory*, *Holographic Universe Theory,* and *String Theory* are but six attempts to explain things like how this book might exist as a one-dimensional string, a two-dimensional brane, or as a quantum superposition of read, returned, *and* overdue.

The best I gather, according to these theories, is that *somewhere* there exists a universe where you've finished reading this book, another where you've understood it completely, and an infinite number of other worlds where you're using my creative work as a coaster for your coffee or as garden compost. The good news is that in at least a few of these worlds, I won a Nobel Prize, a Pulitzer, and was awarded an honorary doctorate in quantum physics.

The Many-Worlds Interpretation

This enticing theory is the stuff of sci-fi writers everywhere. It suggests quantum mechanics describes a multiverse where every possible outcome occurs in different branches of reality. Unlike classical parallel universe theories, these branches continuously split through quantum *decoherence* (we'll revisit this word in the next chapter), creating new realities with each quantum event.

Think of the Many-Worlds Interpretation as a pine forest spreading across a vast landscape, where each tree begets another, and each branch splits into smaller branches in an ever-expanding pattern—except they don't interact. Or imagine the Many Worlds Interpretation as our nervous system, where millions of nerves branch off from the larger trunk of the spinal cord into countless pathways throughout the body.

In both cases, what begins as a unified trunk divides into a multitude of diverging paths—sort of like the way quantum events spawn new branches of a universe. But just as each pine tree grows independently after splitting from its parent, these universe branches also forge entirely separate realities. While they maintain a familial resemblance to their point of origin and share fundamental physical laws of "DNA"—once split—they never again interact—maybe like bubbles floating together in a multi-universe.

But the father of quantum computing, David Deutsch, likely puts it best. His theory suggests that the universe is made up of a network of interconnected qubits, connected by "quantum channels". These channels allow information to be exchanged between them. His theory also incorporates the concept of parallel universes, in which multiple versions of reality exist simultaneously.

Furthermore, the pattern of multiple worlds appears fascinatingly fractal—nature's way of building complexity, except with a crucial difference: unlike true fractals, which maintain perfect self-similarity, each new branch of reality charts its own course through a space of possibility.

This interpretation, also known as the *multiverse hypothesis,* is a fascinating model first proposed by physicist Hugh Everett III in 1957. At its core, it suggests that every possible alternate history and future is physically realized in a different "world" or universe—and that all outcomes are possible and occur in differing (noninteracting) branches of the universe. The implications mean that every quantum decision/measurement creates new branch universes, the total number of universes is constantly growing, there are countless versions of each person living slightly or dramatically different lives, and the total "multiverse" encompasses all possible outcomes of all possible events.

The many-worlds interpretation resolves some paradoxes in quantum mechanics, like the famous "measurement problem" we'll visit in a bit, but it comes with legitimate challenges. For instance, if every possibility is realized somewhere, what does *probability* really mean? And how should we think about personal identity across multiple branches of universes?

While many physicists take this interpretation seriously, it's also controversial. Other interpretations of quantum mechanics, like the *Copenhagen interpretation* or *pilot wave theory*, offer different but equally interesting explanations for quantum phenomena.

Currently, there's no experimental way to prove in the public domain which, if any of these interpretations is correct, as they all make virtually the same predictions about observable phenomena. It's more realistic, I feel, to say that each possesses elements of potential truths. And anything technical or quantum in the public domain, is likely used as an obsolete *front,* or presented as a solvable puzzle—but with purposely omitted pieces.

String Theory

This theoretical framework suggests all matter and forces in the universe are made up of tiny one-dimensional "strings" that vibrate in different ways. These vibrations in multiple dimensions (ten or eleven) potentially give rise to all the particles and forces we observe. While string theory suggests the possibility of multiple universes, this isn't its focus.

Imagine String Theory and the Many-Worlds Interpretation as two different ways multiple universes might exist. In string theory, other universes would be like different field games, each with its own unique rulebook—where in one universe light travels at a slower speed, or in another, gravity works upside down.

Alternatively, the Many-Worlds Interpretation is more like watching the same game play out in a multitude of different ways. It suggests that every time something happens in our universe, new copies split off where things happen somewhere else differently—like one universe where you win the lottery and live happily ever after, and another universe where you spend all your winnings on Spam and end up begging for spare change.

Consequently, in recent years and with the advent of new theories, string theory has fallen out of favor with many (but not all) physicists due to a lack of experimental evidence. However, this theory is still valuable. Even critics acknowledge its worth as a mathematical framework that has led to important insights in theoretical physics and mathematics. Unfortunately, it is not the golden goose of a unified theory of everything.

M-Theory

This theory attempts to unify all versions of string theory into a single framework. It suggests there are eleven dimensions and that fundamental objects called "branes" (membranes of various dimensions) are the building blocks of reality. While mathematically elegant, M-Theory too, remains experimentally unproven.

It is a theoretical framework in physics that attempts to unify all fundamental forces of nature, including gravity, into a single coherent mathematical model. It emerged in the mid-1990s as a synthesis of five different string theories that were previously thought to be distinct.

For ease of reading, let's examine a few aspects of M-theory:

1. It proposes that our universe has eleven dimensions—the familiar four dimensions (three spatial dimensions plus time) and seven additional spatial dimensions that are "compactified" or curled up at microscopic scales.

2. The "M" in M-theory has no official meaning, though some interpret it as standing for "Membrane," "Mystery," "Magic,"

or "Mother" (as in mother of all theories). Edward Witten, who first proposed M-theory, has allegedly deliberately left its meaning ambiguous.

3. It suggests that fundamental particles are not point-like objects, but rather one-dimensional strings and higher-dimensional objects called "branes" (short for membranes).

4. The theory proposes that these strings and branes vibrate in different ways, and these vibrations give rise to all the fundamental particles and forces we observe in nature.

The energy scales required to test many of M-theories predictions are far beyond the capabilities of well-known particle accelerators (but we don't really need them anyway, as we can create much smaller and efficient ones). Some physicists also criticize the theory for being too abstract.

Brane Cosmology

This theory builds on M-Theory, proposing that our universe exists on a 4-dimensional brane within higher-dimensional space. It also offers novel explanations for dark matter and dark energy as effects from neighboring branes.

Brane cosmology emerged from string theory in the late 1990s. The term "brane" comes from "membrane," and proposes that our visible, four-dimensional universe (three space dimensions plus time) is a sort of membrane (or "brane") floating in a higher-dimensional space.

Some concepts of brane cosmology include:

1. The idea that our universe exists on a 3+1-dimensional brane (called a "3-brane"), while gravity can propagate through extra dimensions

2. The existence of other universes hosting branes parallel to ours

3. The proposal that the Big Bang might have resulted from a collision between two branes

4. An explanation for why gravity appears weaker than other fundamental forces (because it "leaks" into extra dimensions)

A famous model in brane cosmology is the Randall-Sundrum model, proposed by Lisa Randall and Raman Sundrum in 1999. It suggests that our universe is a brane embedded in a five-dimensional space called the "bulk," with a warped geometry that could explain the 'hierarchy problem' in particle physics.

The hierarchy problem in physics refers to the large discrepancy between the weak force and gravity, specifically, why gravity is so much weaker than the other fundamental forces. We'll look at gravity (the byproduct of electricity) in a bit.

All universes considered, this theory is highly speculative and involves rather obscure concepts. While considered a promising avenue to unify gravity with other fundamental forces, especially within the framework of string theory and M-theory, it too, like most other aspects of physics, lacks direct experimental verification.

The ADD (Arkani-Hamed, Dimopoulos, and Dvali) Model suggests large extra dimensions where gravity's apparent weakness is explained by its dilution across these additional spatial dimensions.

Butting up against the spiritual again, the Randall-Sundrum model, with its warped geometry, hosts interesting parallels with sacred geometric concepts of nested realities and dimensional hierarchies. Comparatively, the ADD model, with its large extra dimensions, aligns somewhat with sacred geometric concepts of hidden dimensionality and the idea that fundamental patterns repeat across scales. I'll return to and further address some aspects of sacred geometry as we glide along here.

Admittedly, I feel that well-meaning and highly intelligent physicists are trying too hard to squeeze quantum theory into an empirical glass slipper—one that shows the dispersed wave energy of subatomic cracks.

GRW (Ghirardi-Rimini-Weber) Theory

This theory takes a different approach to physics, attempting to solve quantum measurement problems through spontaneous wave function collapse. It proposes quantum systems randomly collapse according to precise mathematical rules, explaining why we don't see quantum superpositions at macroscopic scales.

GRW theory (Ghirardi-Rimini-Weber theory) is an interpretation of quantum mechanics proposed in 1985 by physicists Giancarlo Ghirardi, Alberto Rimini, and Tullio Weber. It attempts to solve the measurement problem in quantum mechanics by modifying the Schrödinger equation to include spontaneous random "collapses" of quantum wavefunctions.

GRW theory proposes that quantum systems normally evolve according to the standard Schrödinger equation and that at random times, particles undergo spontaneous "localizations" or *collapses* that reduce quantum superpositions to more definite classical-like states. These fundamental constants mean that collapses happen very rarely for individual particles but very frequently for large objects (due to the number of particles involved).

GRW theory is significant because it allegedly provides a potential solution to the 'measurement problem' by introducing spontaneous collapses of the wave function without requiring observers or conscious-ness, unlike the Copenhagen interpretation.

If I'm getting this correctly, GRW is considered a "dynamical collapse theory" because it *modifies* the fundamental equations of quantum mechanics rather than just *interpreting* them.

The Holographic Universe Theory

This theory is also known as the holographic principle, another fascinating concept in theoretical physics (and science fiction). The Holographic Principle suggests that three-dimensional reality is an illusion—essentially a projection of information stored on a two-dimensional surface, like how a hologram creates a 3D image from a 2D surface.

The theory emerged from research by physicist Leonard Susskind (and others) in the 1990s. It was built on earlier work by Stephen Hawking and Jacob Bekenstein regarding black holes. Key insights came from studying how information might be preserved and stored on the surface of a black hole's event horizon—rather than obliter-ated inside of one.

This theory helps resolve some paradoxes in physics, particularly regarding black holes and quantum mechanics. The elegant math-ematics describing the surface of a black hole might also be applied to any region of space, suggesting that the entire universe (and its related multiverses) functions like one huge cosmic hologram.

At first it might appear that such a potentially frightening theory might mean that life and our experiences aren't *real*. However, it suggests that there's a fundamental relationship between the way information is stored, and how we *perceive* reality. Life is *real* to sentient beings insofar as we have consciousness, working senses within a biological container, and *believe* it to be so.

The holographic principle has become particularly important in quantum gravity research. It represents one of physics' attempts to reconcile quantum mechanics with Einstein's theory of gravity, offering potential scaffolding for understanding how space, time, and information are fundamentally connected.

Among these five theories, **M-Theory** appears most promising for unification, but remains highly speculative in mainstream science. And we can ask interesting questions about each proposed theory to further propel both our minds and research:

- **M-Theory**: Can we experimentally detect the extra spatial dimensions predicted by this unified theory of quantum gravity? Is there a unique vacuum solution among the vast landscape of possible configurations that M-Theory allows?
- **Brane Cosmology**: How do we test for other branes? Can it quantitatively explain cosmic observations?
- **GRW Theory**: What causes collapse events? Can we measure the precise collapse parameters it predicts?

As we probe beyond the deepest structures of reality that we allegedly can't test, precisely measure, or understand, these theories, much to the consternation of strict material scientists everywhere, echo ancient spiritual wisdom in astonishing ways.

For example, M-Theory's hidden dimensions mirror mystical teachings the globe over, about unseen realms. The Many-Worlds' infinite branches evoke sacred concepts of infinite possibility and parallel realities. Brane Cosmology's higher planes resonate with religious cosmologies describing multiple levels of existence, and GRW's mysterious collapses reflect age-old questions about consciousness and manifestation.

When we get to the intersection of spirituality and science later in this book, we'll explore in more detail: wormholes, free energy, time, remote viewing, ESP, astral travel, life after death, parallel dimensions, God, consciousness, otherworldly beings, secret societies, ancient wisdom, and how all of this and more interconnects with quantum physics.

Taken together, these constructs suggest that ordinary human perception glimpses only a fraction of a vastly enriched reality—one where the infinite unfolds through hidden dimensions, parallel worlds, and higher planes of existence—much like spirituality. And highly curiously, many of the scientists involved in quantum or nuclear clandestine research also tend to espouse and regularly practice different forms of ancient wisdom traditions.

Da Vinci Revisited

As I reflect, I am again reminded of, and inspired by, Leonardo da Vinci. I consider his exhaustive quest to capture atmospheric effects and spatial depth in his paintings. It was his lifelong endeavor, which led to significant innovations in a host of fields. Da Vinci's main challenge as a painter was depicting air and how humans perceive distance and form—what he called "prospettiva aerea" (aerial perspective). My quest here, and that of theoretical physicists, is similarly lofty in its compulsive desire for consciousness to burst forth from the substrate, rise through the subatomic, and find the sun (or Source).

To overcome his challenges, da Vinci studied the way particles in the air affected the appearance of distant objects. He observed that atmospheric haze made distant objects appear progressively bluer and less distinct and developed a subtle gradation of tone and color that created a smoke-like haziness at the edges of forms. He also layered extremely thin glazes of paint to create luminous effects, allowing light to penetrate and reflect off different layers.

Da Vinci enhanced spatial depth through what he called "prospettiva del perdimento" (perspective of disappearance), progressively reducing the clarity of outlines and details as objects receded into space.

Smoke-like haziness and dematerialization are where I find myself when I'm feeling highly spiritual or studying quantum physics—I can almost make out or *feel* the forms, figures, angels, air, energy, and mathematics—until the Source snaps its portals shut, and they elude me. This, I feel, is where da Vinci far exceeded most humans. He had a natural talent for transcending into subatomic worlds to capture phantom landscapes and ghostly forms that he immortalized upon canvasses and clay—the way I hope to do by putting words to paper.

Like his paintings, we see a multitude of colors and motion, which draw out our perceptions—one scene laid out under one Light. Laying his hand to canvas with the stroke of a brush, da Vinci was Source made manifest, inhabiting human form to create.

What makes da Vinci's achievements particularly remarkable is that he arrived at his solutions for capturing wavelengths and subatomic particles through detailed scientific observation combined with insatiable curiosity, and the courage to ask questions. He also experienced visions. He was a master at capturing *perception* and recognized something about other dimensions that most of us miss because he gave detailed attention to energy's virtually invisible minutia.

Da Vinci didn't just approach the fundamentally mysterious nature of reality, he immersed himself. Like quantum physics, he recognized that solid matter is not as solid as it appears. And similar to ancient mystics and modern physicists, he sought to reveal the subtle interconnections between all things. His paintings, drawings, and sculptures are meditations on the nature of reality itself—both highly technical and deeply spiritual.

The integration in da Vinci's work reminds me of David Bohm's "implicate order"—the deep interconnectedness of all things.

CHAPTER THREE

Coherence, Decohrence, and Chaos, Oh My!

Coherence, in its basic form, refers to a *logical, consistent* inter-connection between parts that form a harmonious whole. For example, a computer typically requires collaborative hardware, a human, and chips, to continually and efficiently work. Leave out or break a component, coherence is disrupted.

Coherence meanders through a multitude of minds, dimensions, and systems. In physics, coherence describes waves that are in phase and work together. In writing, coherence refers to ideas that logically flow and meaningfully connect. I could go on...

Coherence takes many forms through a variety of sentience on this planet. For example, from a biological perspective, coherence might be viewed as the remarkable coordination between different systems in living organisms, where cells, organs, and neural networks each have a different job but work together to maintain life.

Coherence, as applied to consciousness, attempts to describe how our individual minds create a unified, consistent experience of reality apart from some hodge-podge of disparate sensory inputs or fragmented memories infringing upon sensible everyday function. It is, in other words, how we make *sense* of the world around and within us. But even this poses a quandary, which some neuroscientists call the "binding problem"—how does our brain create one *coherent* stream of consciousness from a myriad of crisscrossing and parallel processes?

Or should we look at this instead, as a *cloud* of pathways? In the quantum world, cloudy pathways aren't always linear trails but clockwise and anti-clockwise vortexes that unfurl when we settle on a direction—but the direction isn't always clear. It's more of a labyrinth—or a lattice.

For example, if I point to a house and ask a group of people what it is, we would all likely agree, based on coherence, that what we're seeing is a house. We may have many alternative and subjective *perceptions* and interpretations regarding the house but it's still a house. This isn't particularly interesting and is considered *low information* by mathematical standards, until we realize how billions of neurons communicate and synchronize with individualized subjective experience to provide a perception of *unified* consciousness among a group of people.

We all seem to collapse on the same basic structure of a house and call it thus—even if we might disagree on its meaning, style, structure, context, or contents. This tells me that somewhere, conscious energy is exchanged *outside* of a warm, wet lump of fat, salt, protein and carbohydrates we call a brain. But what sort of energy or information is transferred and from where does it originally arise?

What interests me about coherence is how it suggests that consciousness might not be just about having awareness, but about having integrated, *meaningful* awareness—a corner piece of a confounding puzzle that is only a necessary part of an infinitely larger and inter-connected whole.

Curiously, philosopher Daniel Dennett suggested that conscious-ness might be a sort of "user illusion", a *fiction* our brains use to tell a coherent story about all the parallel processing happening within, around, and outside of us. Consciousness, it appears, is what allows humans to *perceive* and *feel.*

Dennett's quote sounds eerily like fiction writing and the *interface model of physics*, which posits that our understanding of the universe is shaped by the tools we use, the sensory and cognitive limits we possess, and the models we construct. It highlights that physics is as much about human interpretation as it is about uncovering objective reality and measuring matter.

Swinging over to the wilder side of this pendulum we have Itzhak Bentov, who helped launch the biomedical engineering field with his invention of the cardiac catheter. Bentov believed that human bodies mirror the universe down to the inner workings of each cell. He claimed that humans are pulsating beings in a vibrating universe, one of constant motion between the finite and infinite, where all matter is conscious, and our universe holographic (a projected image). Some researchers have referred to this as *Life Physics* or Biophysics. Governments have been studying aspects of these ideas for decades through ongoing clandestine research.

If this is so, and Bentov is right, his theory lends credence to the idea that humans are made in the image of a God—an intelligent consciousness that resides in and around us, one of which we are a part, equal in all respects, within a universe that requires our obser-vances and participation, or dematerializes into a static wave state.

Bentov almost mirrors Stanislov Groff's *out there* but interesting idea that we live in a fractal universe. Groff, a psychiatrist, was one of the pioneers who developed transpersonal psychology. He proposed that human consciousness could access different "levels" of reality, from everyday awareness to profound mystical states.

Interestingly, the National Security Agency (NSA) allegedly studied these levels of reality through a group known as the *Advanced Communications Intelligence Organization*. The CIA mirrored similar research with its Gateway and Stargate programs, among others. One simple google of *MK Ultra* will likely lead you into a world you never knew existed—but fair warning: It is a scary and ethically disappointing place you may not want to visit.

Groff's LSD research, and later his holotropic breathing techniques were used to explore the depths of the human psyche and our relationship to the universe. He believed that each state of consciousness available to humans repeated at different scales, like mathematical fractals.

Moreover, fractals and holograms (holographs) are considered completely different concepts. Yet both can be combined to create fascinating visual phenomena where a hologram displays fractal patterns or structures. This is an interesting intersection (like a 3D-printed Mandelbrot set or a natural fractal like a Romanesco broccoli).

A few artists and researchers have explored this combination. For example, there have been experiments creating holographic displays of 3D fractal patterns like the Sierpinski tetrahedron (a three-dimensional analog of the Sierpinski triangle). This idea was taken from much older 13th century stonework inlays known as *Cosmatesque* in Italy. It is significant because it demonstrates the concept of self-similarity and the fascinating properties of fractals where the surface area remains constant while the volume approaches zero with each iteration.

Maybe consciousness is like a mosaic and we, as its individual pieces, are arranged to form the appearance of a whole—where quantum physics is like a wall holding up and giving light to the Source's creation and experience through the senses, wave fields, and energy of each unique appearing tile.

The interesting challenge with fractal holograms is that true mathematical fractals continue infinitely, while holograms have physical resolution limits. Ultimately, things become grainy if we probe too far into the void of quantum physics—as if we have a consciousness horizon—one with perception limits.

So, any holographic representation of a fractal would be limited to a finite number of iterations of the pattern. Any attempt below a theoretical limit and those annoying quantum effects step in! When we dare try to tread where angels might fly above or below our biological and time-dependent radar, the gods seem to block our path and roll up the vortex!

Quantum coherence is considered a fundamental property. Quantum particles maintain a consistent phase relationship over time—essentially, their wave-like properties remain synchronized yet can interfere with each other. Lasers and superconductors fit here.

QC allows particles to exist in multiple states simultaneously (superposition) until an observer arrives and a collapse takes place that leads to a defined state. It is essential for quantum computing and many other quantum effects but is fragile and easily disrupted by interaction with the environment.

On a human level, quantum coherence improves the efficiency and accuracy of cellular signaling, allowing for precise energy transfer of information to improve biological coordination. In plants, we call this photosynthesis. It is possible, as we'll see, that humans can also bioengineer coherence to optimize mitochondrial efficiency such as increasing endurance, reducing fatigue, and other extraordinary sentient perks such as increased intelligence.

Moreover, human senses, such as vision, hearing, or magnetoreception, benefit from quantum coherence by increasing sensitivity and accuracy. We can slow aging, reduce mutations, and enhance a human body's ability to swiftly recover from illness or injury through targeted cell therapies due in part to a form of quantum coherence—and so much more!

Real-World Examples

- **Lasers:** All photons in a laser beam are coherent and move in sync with aligned wave patterns, making them focused and powerful.
- **Superconductors:** help electrons move coherently through the material and permits electricity to flow with zero resistance but only works at very low temperatures where coherence can be maintained. Cooling reduces vibrations, allowing electrons to move without frequent interruptions and form *Cooper pairs*, which are bound together through interactions with the lattice (phonons). These pairs move in a coherent quantum state, acting as a single unit.
- **Bose-Einstein Condensates:** Atoms are cooled to near absolute zero to become coherent and behave as a single quantum entity, moving in perfect synchronization (applications include quantum computing, gyroscopes, precise atomic clocks, communications, and navigation).

What Destroys Coherence:

- Heat (thermal vibrations)
- Measurement/observation

- Interaction with other particles
- Environmental noise

Quantum computers require extreme cooling to maintain coherence long enough to perform calculations. This sensitivity arises because quantum bits (qubits) rely on delicate quantum states, such as *superposition* and *entanglement*, which is easily disrupted by external factors such as vibrations and decoherence of electromagnetic interference. (*Note:* I've looked at some Tesla experiments that appear to ignore cooling while continually oscillating, especially when frequency is increased, and amplitude is decreased).

Quantum Decoherence

Quantum decoherence attempts to explain how the quantum interacts with an environment to transition or cross-over from the unseen (subatomic) micro-scale to the observable macro scale. It occurs when a quantum system interacts with its surroundings, causing *superposition* to break down and reveal much more comfortable classical states, ones we can measure and observe.

Here's an analogy that might help: Imagine a placid ocean that's as smooth as a mirror on a still day and we can't discern any waves. This doesn't mean that the ocean is suddenly and infinitely void of waves. It means that the ocean has the *potential* to remain calm— or—to make waves. Decoherence causes the ocean to settle into one specific pattern, placid or wavy. I liken it to the precise moment of disruption in a system, which causes change in that system.

It's also important to note that when a quantum system interacts with its environment, information about its quantum state spreads into the surroundings, making the original superposition impossible to recover (you never step into the same river twice).

Decoherence typically happens extremely quickly, especially for larger systems. This is why maintaining quantum states for quantum computers is so challenging—they need to be heavily isolated from their environment. This is also part of the reason why human beings will never be fully controlled. Complex systems are impossible to precisely and continually measure or dominate.

Decoherence, on the other handedness (pun intended), is crucial for understanding the boundary between quantum and classical physics, and it's particularly relevant for practical applications like quantum computing, where preventing decoherence is considered by mainstream science, a major technical challenge. Decoherence also helps explain why we only ever measure definite outcomes in quantum experiments, even though quantum systems can exist in multiple states simultaneously before measurement.

Additionally, we are repeatedly told that our brains, due to its warm, wet environment, is prone to decoherence, and therefore, unable to achieve or maintain quantum states. The latest public research shows us otherwise and says, "hold on a moment". And if a brain can and does bypass decoherence, so can a computer. More on this shortly.

Meanwhile, it is difficult outside of the world of science fiction or a psychedelic trip to comprehend how something like electrons and its alter ego, neutrinos, might exist everywhere at once until we focus attention or observe—and it collapses into something solid under the *weight* of our proverbial glare.

Chaos

Chaos, despite its colloquial meaning, is a specific phenomenon in physics and mathematics—and a bit of a misnomer. It refers to systems that follow fixed rules and are considered deterministic even though they are extremely sensitive to initial conditions. In other words, small changes lead to dramatically different outcomes over time—like weather.

For instance, hurricanes and snowstorms are classic examples of decoherence where tiny variations in initial weather conditions, like barometric pressure or frontal boundaries, often lead to vastly different weather patterns days later—which some of us in the direct path of severe weather might refer to as *chaos* in their aftermaths.

Notably, chaos reveals that unpredictability is inherent in most complex systems. A complex system includes the human brain, ecosystems, weather patterns, social networks, crowds of people, the global economy, power grids, ant colonies, the human body, cities, and the Earth's climate system, to name but a scant few.

Moreover, this unpredictability means complex systems can never be completely controlled despite our worst intentions or best technology. Why? Change happens. However, researchers have recently studied crowd dynamics at the Running of the Bulls festival in Pamplona, Spain to accurately predict crowd movement and prevent danger. Their research suggests that crowd movement *is* predictable when crowd density reaches a certain level, and moves like a fluid, in anticipated time intervals, shifting from random to rhythmic—like flocks of birds or schools of fish.

Alternatively, many people mistakenly believe that the "butterfly effect" and chaos are the exact same. While they are related, the butterfly effect is a metaphor and is an example of chaos phenomena. The butterfly effect refers to small *changes* in initial conditions that might lead to large, unpredictable consequences. Chaos theory encompasses the study of complex systems that exhibit extreme *sensitivity* to initial conditions.

Interestingly, the "butterfly effect" resonates with ancient Eastern philosophical ideas about the fundamental interconnectedness of all things. Many ancient traditions, from Taoism to Indigenous wisdom, emphasize how all parts of nature and consciousness are deeply linked in ways that aren't immediately obvious to those who have their heads buried in technology or strict scientific paradigms.

Additionally, chaos theory reveals how seemingly random phenomena contain deep mathematical patterns and strange attractors. This mirrors what many ancient traditions spoke about regarding a well-hidden underlying principle governing 3-4D reality.

The Hermetic axiom "as above, so below" suggests similar patterns manifesting at different scales, not unlike how fractals mentioned earlier show self-similarity across scales.

Some researchers have proposed that consciousness itself may be an emergent phenomenon arising from complex systems operating at the edge of chaos. This connects to how several ancient traditions viewed consciousness as an inherent property of nature emerging from fundamental patterns that might have manifested as archetypal in humans.

Likewise, the ancient Greek concept of the *logos* as an underlying intelligent principle organizing the cosmos has some interesting parallels with how chaos theory reveals complex organization emerging from simple rules. Much of this thought led to the art of philosophy, the pondering of quantum worlds, and the development of sacred rituals where spirit forces inhabit all of creation.

Yet in the words of Democritus, "Nothing exists except atoms and empty space; everything else is just opinion."

Socrates might counter him, "One thing I know is that I know nothing. This is the source of my wisdom."

This leaves us swimming in a vast illusion of solidity, built from restless atoms dancing through an ocean of empty potential. It means nothing but is everything. The science butts up against the sacred.

Democritus, known as the "laughing philosopher," proposed atomic theory with the idea that all matter consists of invisible, indivisible particles (atoms) moving through a void. His materialist philosophy suggested that understanding the fundamental physical nature of reality could give humans immense power.

Moreover, he eerily foreshadowed the Manhattan Project, where understanding atomic structure led to unprecedented destructive capability. His view that humans could master nature through knowledge proved both prophetic and perilous.

Socrates, conversely, focused on ethics and human wisdom. His famous statement emphasizes questioning assumptions and deeply connects to the moral dilemmas faced by atomic scientists.

Many Manhattan Project physicists, like J. Robert Oppenheimer, wrestled with Socratic-style ethical questions about whether humans possessed the wisdom to wield such knowledge responsibly.

Oppenheimer famously quoted the *Bhagavad Gita* after the first atomic bombing test: "Now I am become Death, the destroyer of worlds"—a very Socratic recognition of the gap between technical advancement and moral wisdom.

The synthesis of these philosophies illuminates a core tension of the quantum age: Democritus showed we could split the atom, while Socrates asked if we should.

CHAPTER FOUR

Unorthodox Physics, Death, and the Disavowed Realm of Quantum Potentiality

J. Robert Oppenheimer is considered the father of the atomic bomb. He is credited for leading the Manhattan Project at Los Alamos laboratory during World War II. He was also chairman of the United States Atomic Energy Commission, today known as the Department of Energy. Oppenheimer was a wildly intelligent and complex man. The child of non-practicing German-Jewish parents, Oppenheimer grew up in relative material comfort and attended boarding school.

Through a vortex-like turn of events that spin through more complex systems than these pages allow, Oppenheimer was both celebrated and vilified—celebrated for his out-of-this world intellect and scientific advancements—and vilified as a communist sympathizer. He was also once accused of being a Russian operative. His high security clearances were revoked and then later reinstated. He worked with a group of German scientists, some former Nazi's, whose backgrounds were whitewashed under *Operation Paperclip*. I often wonder how he felt about that, and something in his eyes, revealed through pictures, captures this swirling dichotomy of love, charisma, and compassion juxtaposed against gloom and an undercurrent of something far more sinister. The man literally carried the survival and destruction of humanity and the planet on his shoulders.

Oppenheimer eventually retired to the US Virgin Islands where he spent his days sailing with his daughter Katherine (Oppenheimer also maintained a New Mexico ranch not far from Los Alamos, where he worked.) After his death, his daughter inherited the island property. She'd obtained a job working as a translator at the United Nations but was denied a security clearance by the FBI based on her father's years-old charges. She died of hanging in 1977 at the beach house, her death ruled a suicide. The Oppenheimer home was given as a gift to the people of the island but was later destroyed by a hurricane.

While intriguing, most of the information about Oppenheimer omits or barely mentions his relationship with Dr. David Bohm, an exceptionally bright graduate student of Oppenheimer's at the University of California, Berkeley, who was considered one of the greatest theoretical physicists of his time.

In fact, Bohm conducted pioneering work on plasmas at what was once the Lawrence Berkeley Radiation Laboratory. He discovered that electrons, once stripped of atoms, do not act like separate particles but as elements of an organized whole, as if some organic and arguably conscious process was organizing and directing this behavior.

Bohm's early life was far less privileged than Oppenheimer's. Bohm was the son of a used furniture salesman in Wilkes-Barre Pennsylvania, raised around Midwestern coal mining, auto works, and steel. He was forged from the backbone of small town hard-working, God-fearing people, who spent their lives digging ditches to ensure the wealth of others. Bohm was a man whom Oppenheimer took under his wing before he threw him under a bus...

The Weak Interaction

As a supplement to the above story, I briefly introduce the fundamental interactions of nuclear and particle physics: The *weak interaction*, also called the weak force.

It is one interaction that also includes *electromagnetism*, the *strong interaction*, and *gravitation*. These interactions appear in nature and are not reducible to more basic interactions. It is the mechanism of interaction between subatomic particles that is responsible for the radioactive decay of atoms, which is, essentially, energy. This energy, in the form of ionizing radiation, contains significant energy levels.

Moreover, the weak force is the only force that violates parity symmetry (it treats left-handed and right-handed particles differently) and is stronger than gravity but much weaker than both the electromagnetic and strong forces at typical particle physics energies.

The parity symmetry violation by the weak force is one of the most fascinating features of particle physics. Parity symmetry is the idea that the laws of physics should be the same when viewed in a mirror.

Imagine you're watching a physics experiment and its mirror image. If parity is conserved, both should follow the same rules. However, the weak force breaks this symmetry in a substantial way by interacting with particles based on their chirality (ability to change form):

- The weak force *ONLY* interacts with left-handed particles and right-handed *antiparticles*
- It completely ignores right-handed particles and left-handed antiparticles

- This preference for one handedness (a particle's spin and direction of motion) over another is what is meant by parity violation

A famous experiment demonstrating this chirality was conducted by Chien-Shiung Wu in 1956 (the "Wu experiment"). She observed the beta decay of cobalt-60 atoms aligned in a magnetic field.

In a mirror-symmetric world, these electrons should have been emitted equally in all directions but instead, were preferentially emitted in one direction relative to the nuclear spin. Her work definitively showed that parity symmetry was violated in weak interactions.

Wu's discovery was revolutionary. It was the first time physicists found a fundamental force that distinguished between left and right. It's as if the universe has a built-in preference for "handedness" in weak interactions.

Think of it like this: if someone watched a recording of a weak interaction and its mirror image, they could easily tell which was the real interaction and which was the anti-image (for understanding, think of the anti-image as a reflection—it's sort of *real* too, but only as an image of the true thing). In electromagnetic or strong interactions, we can't differentiate or locate the point of separation between a real/mirror image.

The fact that such a fundamental symmetry is violated in the *weak force* tells us something deep about the universe—that it has a preferred "handedness" built into its most basic laws. This violation shows up in many subtle ways in particle physics and may be connected to the matter-antimatter asymmetry we observe in the universe. The weak force could theoretically play a role in *quantum coherence* mentioned in the previous chapter.

While this asymmetry allowed consciousness to emerge (by allowing matter to exist), it's arguably less clear if it plays an active role in consciousness itself—this is what we've been taught—to doubt asymmetry plays a role in consciousness—something akin to dark matter and energy. I find it odd that many scientists accept dark matter and energy as real and existing 'out there' for their research purposes but cannot exactly pinpoint their location or concretely define either—much like consciousness.

The difference mostly lies in the discussion. Scientists readily make room for invisible phenomena like dark matter and dark energy—despite never having directly observed or accurately measured them. Yet they'll dismiss anecdotal experiences of other supernatural phenomena as complete bullshit, much like how competing religions often condemn one another as ignorant, evil, misguided, or wrong.

And they must—because in their respective worlds of science and religion, to do otherwise invites charges of pseudoscience or sin. Whether it's a lab coat or a pulpit, stepping outside the accepted dogma can get you laughed out of a room—or heavily judged and pitchforked out of a promising career in soul-saving.

Now, let's return to the relatively unknown coherent-decoherent and terminally chaotic relationship between Oppenheimer and Bohm. But first it's important to receive a crash course (if you're new to all of this) or a refresher (if you're a scientist type scoffing over this book), about Einstein's *spooky action at a distance,* something also known as *nonlocality,* which is different from superposition.

Nonlocality in quantum physics refers to the ability of particles to instantaneously influence each other regardless of the distance between them. Broken down, nonlocality challenges both intuitive and materialistic understanding of reality as we think we know it, in a few ways.

In classical physics, objects can only affect each other through local interactions—meaning direct contact or forces that travel through space at finite speeds (like gravity or electromagnetism). We naturally expect that what happens in one place shouldn't instantaneously affect something far, far away—but it does!

For example, if I suddenly die alone, my spouse who's out fishing would not typically be instantaneously impacted (yes, we could and will argue this later). In a 3-4D, strictly material and empirical world, my husband would suffer affect when someone else makes direct contact with him, using energy in the form of a phone call or physical visit to relay information about my death. The human relaying the message must use time and space to deliver this tragic news—and is usually bound by such.

Alternatively, quantum mechanics revealed that two or more particles can become "entangled," meaning their quantum states are fundamentally interconnected. When one particle is measured, information about the other particle is instantly known, if they are *entangled*, even if they're separated by vast distances—which might be like my husband's consciousness having, somewhere, been calibrated to my consciousness, where we developed a sort of interconnected consciousness through an adult lifetime of sharing DNA and mental energy exchanges.

Depending upon my husband's level of *awareness,* he may in fact, at my moment of sudden death, *perceive* that something is very wrong, even if he lacks the exact information or is unable to describe his abrupt onset of doom and gloom.

Taking a quantum and supernatural leap here, some anecdotal experiences include mothers whose children have *appeared* to them at death to share that they are okay. As a mother of six grown children, I view this phenomenon as science we don't yet understand and entirely possible—perhaps based in DNA, familial, energetic, or some sort of blood entanglement (which we'll also examine further from a credible research standpoint).

Given the lack of concrete mechanics, definitions, and measurement in quantum physics—especially regarding consciousness, which remains as enigmatic as dark energy and dark matter—one might speculate that post-death consciousness emits its own transferable halo of energy, forming a kind of wireless connection or wave field to those left behind.

Since matter, we're told, is primarily influenced by gravity (perhaps even metaphorically, in certain contexts), this idea raises intriguing questions about the nature of consciousness beyond physical existence—where gravity *feels* to be more of a byproduct of electrical fields and waves of graininess.

Additionally, nonlocality seems to violate special relativity's principle that nothing, including information, travels faster than light. However, I'm reminded by engineers, that while the correlation between entangled particles is instantaneous, nonlocality can't be used to *transmit* actual information faster than light. So, what then, is transmitting thought? Axions? Neutrinos? Or do thoughts reside or originate outside of 3-4D space and time?

On this last point, Dr. Edgar Mitchell, would agree that thoughts and consciousness travel faster than the speed of light and even once, during an interview, light-heartedly apologized to the dead Albert Einstein for thinking him wrong regarding relativity.

This research on Mitchell led me to ask: What if consciousness is *entangled* or superpositioned with everything? No transmission of actual information would be required because it would already be anywhere and everywhere, ripe for proverbial quantum picking as needed—like the Akashic Field—and completely whole—like Bohm's theories of interconnectivity.

Notably, Mitchell was a United States Navy officer and aviator, test pilot, aeronautical engineer, ufologist, and NASA astronaut. As the Lunar Module Pilot of Apollo 14 in 1971, he spent nine hours working on the lunar surface in the Fra Mauro Highlands region and was the sixth person to walk on the Moon.

Upon his return to Earth, he experienced what might be described as a life-changing transcendent state—an "explosion of awareness". He wasn't the only astronaut to realize this mystical transformation. A myriad of astronauts return from space somehow spiritually charged. Many describe their off-Earth voyages as life-changing, which fundamentally alters their worldview and priorities. The phenomenon is called the "*Overview Effect*," and it's a well-documented psychological and cognitive shift in consciousness astronauts experience when viewing Earth from space.

Perhaps seeing Earth through a tiny window, a lone oasis in the vast darkness of space, leads astronauts to realize the special uniqueness of our planet. Some report the silliness of artificial national boundaries and a greater understanding that humanity and all life on Earth is fundamentally and universally connected. Many astronauts also return with a heightened concern for environmental issues and sustainability—and frequently report a reduction in ego-based thinking.

Frank White, who coined the term "*Overview Effect*," has extensively documented these experiences. While the experience isn't considered *supernatural*, it represents a major shift that comes from being one of the few humans on Earth to see our planet from a unique vantage point.

Additionally, Mitchell conducted, unbeknownst to NASA officials, unsanctioned extra-sensory-perception (ESP) experiments with his friends back on Earth while he was in space. Focusing on a specific Zener card at a certain agreed upon time, he sent an intention of the card to his friends on Earth. The results (his *hits)*, proved far above chance, scoring fifty-one out of two-hundred correct. He was so moved by his deeply personal experience in space, and its life-altering impact, that he dedicated his life to the study of consciousness and founded the Institute of Noetic Sciences.

According to Mitchell, quantum nonlocality suggests that signals or information may indeed travel faster than the speed of light. However, he later tempered this assertion for the sake of scientific rigor, acknowledging that this remains a complex and unresolved area—even though quantum nonlocality has been repeatedly demonstrated in laboratory settings. His trepidation was understandable.

Like the concept of infinity, nonlocality is unsettling to scientists steeped in the Cartesian model, largely because theories such as Bohm's quantum potential or anything faster than the speed of light appears to lack any identifiable physical source. This ambiguity often leads to a quick retreat from ideas that can't be lab tested.

Similarly, Mitchell viewed nonlocality as a valid concept because quantum relationships seem instantaneous for an infinite space. His problem, he noted, was the idea of infinity, which humans can't grasp. He felt that *infinite* is big, but nonlocality as infinite can't be scientifically demonstrated, even if it's what the entire theoretical structure shows.

This poses a great nonsensical question: Is nonlocality infinite? It appears that our very own *physicality* gets in the way of our ability to test for a definitive answer. And this, Bohm thought, endangered knowledge, which would be at risk of becoming static—because the process of thought is merely a mental representation we assign, often incorrectly, or in limited ability, to our world. In Bohm's mind, humans view thought as separate and independent of its content, which is an artificial construct. Limited content excludes the whole of reality—much like science and religion.

Returning to a classic example of nonlocality described earlier in this book, we can measure the spin of entangled electrons. If you measure one electron's spin as "up," its entangled partner will instantaneously be in the "down" state, regardless of how far apart they are, these being the only two options publicly available. This has been experimentally verified numerous times, most notably in experiments by Alain Aspect in 1982.

But here's the twist: The electrons are not *"up"* or *"down"* in a simple, everyday sense. Instead, they're a fuzzy mix of both. It's like they agreed to their states in advance without deciding who's who until the moment we look—so, are they then really two electrons— or one? Are we really separate humans living unique lives, or one universe expressing itself in a single folding and enfolding wave that only resembles or presents individual parts to our limited senses?

The implications of nonlocality are both substantial and enigmatic. It suggests that at a fundamental level, the universe isn't composed of separate, independent parts, but is rather an interconnected whole where distant objects have immediate, potentially instantaneous connections. This challenges classical knowledge regarding locality, causality, and the nature of space itself. Maybe nonlocality requires no transmission, existing outside of time and space, and everywhere at once.

Nonlocality remains one of the most puzzling aspects of quantum mechanics, and its philosophical implications will likely continue to be debated long after most of us are biologically dead. While humans can mathematically describe and experimentally verify nonlocality, again, there's still no consensus on what it truly means for our understanding of an illusion passing as reality in a 3-4D world.

Alternatively, the relationship between nonlocality and entanglement was historically important. Einstein, Podolsky, and Rosen (EPR) first highlighted entanglement in their famous 1935 paper, *"Can Quantum-Mechanical Description of Physical Reality Be Considered Complete?"* They used it to argue that quantum mechanics must be incomplete because it seemed to allow for "spooky action at a distance" (nonlocality).

Later, John Stewart Bell's mathematical theorem showed that no local hidden variable theory could reproduce all the predictions of quantum mechanics, proving that nonlocality is an inherent feature of quantum mechanics.

Now, let's get back to Bohm.

In 1952, Bohm published, *"A Suggested Interpretation of the Quantum Theory in Terms of 'Hidden' Variables"* in *Physical Review*. This was Bohm's paper presenting what became known as the de Broglie-Bohm theory or Bohmian mechanics—a deterministic interpretation of quantum mechanics that attempted to resolve some of the paradoxes of The Copenhagen interpretation, developed primarily by Niels Bohr and Werner Heisenberg in the 1920s. Bohm introduced the concept of "hidden variables."

Bohm's research on the "quantum potential" is a key component of his "de Broglie-Bohm theory" (also called Bohmian mechanics), which interprets quantum mechanics by proposing that particles have definite positions guided by a wave function, with the quantum potential acting as a force-like influence (a propulsion or current) that directs their motion, effectively providing a causal explanation for quantum phenomena where traditional interpretations appear probabilistic. But what sets it in motion?

Bohm's quantum potential simplified (with partial credit here to AI):

- **Derivation from the wave function:** The quantum potential is mathematically derived directly from the wave function of a system, meaning it is *completely* determined by the quantum state.

- **Non-local nature:** Unlike classical potentials, the quantum potential can be **non-local**, meaning it can influence a particle's behavior even when separated from the source of the potential by a large distance.

- **Explaining interference patterns:** Astoundingly, by incorporating quantum potential into classical equations of motion, Bohm's theory explains the interference patterns observed in double-slit experiments, where particles behave as waves due to the guiding influence of the potential. And this largely went ignored for decades! How could such an extraordinary man and his research be pushed into such an unobservable realm?

Even more intriguing, the Bohm-Oppenheimer relationship is fascinating because Oppenheimer had initially supported Bohm, hiring him at Berkeley and highly praising Bohm's early work in physics. And once Oppenheimer was offered a position on the *Manhattan Project*, he initially brought Bohm into the fold at Los Alamos. Oppenheimer basically treated Bohm like a son and touted him as one of the greatest minds in physics—until he didn't.

What few people know is that it was Bohm who supplied a host of mathematical equations for the nuclear bomb. You can read that last sentence as many times as you'd like and I'll stand by it—remember, as I said in the intro—this book like my life is a fiction. I also realize that people have been thrown out of windows or strangled with their own catheters under the fiction of suicide, for, well, telling fiction.

Moving along, when a US Army general running the Manhattan project at Los Alamos told Oppenheimer to dispense with Bohm because he didn't have and wouldn't receive the secret clearances required for the job, Oppenheimer was forced to dismiss Bohm. When Bohm lamented the fact that he wouldn't be able to complete his doctoral thesis (because Oppenheimer was his doctoral advisor)—a deal was reached. Oppenheimer took and used Bohm's mathematical equations and research in exchange for giving Bohm a PhD. True fiction.

Afterward, the relationship soured.

Unfortunately, McCarthyism (late 1940'-1950's), accused many Jews of being secret communists, including Oppenheimer, who was a non-practicing Jew and a quasi-Hinduist in his later years. Bohm too, had a spiritual leader in J. Krishnamurti—searching far beyond physics (or nearer than we've realized) for answers to the cosmos and consciousness.

Ultimately, Oppenheimer was forced to testify under oath and to name names of other suspected communists, likely to save himself. He complied, implicating a host of highly intelligent cohorts and colleagues, including Bohm. Promising careers and lives were upended and ruined.

However, Oppenheimer didn't get away unscathed. As I mentioned before, he too, was arrested, stripped of his clearances, and faced an inquisition before demanding a trial. He was ultimately exonerated. Later, it was discovered that he *had* sent monthly financial support to one communist group and likely supported a few others.

Bohm returned to Princeton and just when he was in line to become Einstein's personal assistant, after having reluctantly left Los Alamos—Bohm was subsequently and unceremoniously fired. He was forced to leave the country and take a new position at the University of Sao Paulo in Brazil before moving on to the Technion in Haifa, Israel.

After this, Bohm wound up in London making another contribution to quantum physics with the Aharonov-Bohm effect, which showed that an isolated line of magnetic force can effect electrons that pass around it without encountering it. Oppenheimer privately bitched Bohm out for not having left the US sooner.

While in Brazil, Bohm wrote his revolutionary paper about the hidden variables of quantum theory. Students of physics back in the United States were riveted. But the paper received a cricket-like reception from Bohm's peers around the globe. When one student presented Bohm's paper to Oppenheimer at Princeton, and asked what was wrong with it, Oppenheimer allegedly muttered, "*Nothing*."

Many leading physicists and promising post-doctoral students were then gathered by Oppenheimer, and following his influential lead, and likely fearful of the career-ending power he wielded, dismissed Bohm's *hidden variables* theory as nothing more than a regression to classical determinism. The question is: Who pushed Oppenheimer into doing this and why? Those memos remain classified.

This organized resistance to Bohm's work had little to do with science and much more to with politics, clandestine projects, and secret money. *Hidden Variables* likely would have upended secret quantum physics research, the financial incentives of several covert operatives, and was therefore touted, if brought up, as a deeply flawed philosophy existing on the fringe of science. Parts of it were also redacted and a whitewashed version later appeared in the public domain—the original was either destroyed, "lost", or classified.

Consequently, Neils Bohr's *Copenhagen* interpretation became the deeply entrenched quantum studies standard, and challenging it was viewed as practically heretical by most physicists and government contractors—the way religious leaders might defend to the death that *their* God is the *right* one above all others—even though we're all likely worshipping the exact same God who appears in a myriad of forms and energies—but is really One (however defined).

Consequently, the Copenhagen interpretation states that a quantum particle exists in all possible states simultaneously until observed, at which point it collapses into a single, definite state based on the measurement. Again, hubris, arrogance, and ego got it only partially right within a forcibly constrained but highly imaginative and publicly promoted scientific model.

Like religion, this is where fear and conformity arrived to rule the plot. This chaotic drama dismissed reason in favor of maintaining the static gravitational and rudimentary materialist status quo we have today. In this case, no questions, alternative theories, holomovements, or energies allowed—and—all anomalies will be confiscated, classified, or thrown in the trash. It was, at the time, and currently, every man or woman for him or herself in the realm of covert nuclear or quantum physics (along with classified mathematics and information theory).

Ironically, Bohm's interpretation has gained more respect in recent decades, with some physicists viewing it as an important interpretation of quantum mechanics, particularly for understanding phenomena like entanglement.

Ultimately, the *Implicate Order* Bohm proposed, which was briefly touched upon in an earlier chapter, claims that the universe we observe (the "explicate order") is just the surface manifestation of a deeper, hidden reality (the "implicate order"). He used a memorable analogy: Imagine a cylinder containing glycerin, and when you place a drop of ink in it and turn the cylinder, the ink appears to spread out and disappear. Turn it backwards, and the drop reforms. The spread-out ink represents the implicate order, while the visible drop is the explicate order.

Bohm saw reality as a constant and interconnected flow, what he termed the "holomovement." Rather than seeing things as separate objects, he viewed everything as part of an undivided whole in constant motion—or giving off the illusion of constant motion to sentient beings trapped in a 3-4D world.

Like a hologram, where each piece or pixel contains information about the whole image, Bohm believed every part of the universe contains information about the entire universe—a bit like a fractal but not exactly—maybe more like a mosaic—and only *appears* to be self-contained as independent or separate. Additionally, he claimed that language deceives us about the nature of reality.

We generally consider language (if we consider it at all) to be a natural way of communication—conveying ideas or direction that won't limit or confine our worldviews to any specific framework.

Bohm, disagreed.

Through his work he shows that language applies extraordinarily strong and subtle pressure on human beings to view our world as separate, disjointed, or disconnected—despite advancing technology. This led him to create a little-known language mode he termed *rheomode.* The rheomode emphasizes verbs over nouns. It focuses on action, transformation, and process rather than static objects—and reflects his "implicate order" where everything is in a state of constant flux.

Basically, Bohm thought that by changing language, we change thinking, helping us move beyond the limits of classical, mechanistic thinking into a whole and interconnected mode of *being*.

Circling back to the universe-as-a-hologram model, this line of thought is initially disturbing because it lends credence to an idea that everything under the sun, and beyond, is a simulation, a fragmented hologram, where something deeply intelligent brings everything into existence and we are illusory reflections of it or powerless subjects without free will. You can bet I'll return to that latter topic much later in this book because my primal-based ego and pre-programmed faith finds it too disturbing to accept that we have no free will—a thought, I'm sure, that has secretly crossed and vexed the minds of secret ops officials, scientists, politicians, and religious leaders everywhere.

Similarly, Bohm's interpretation of quantum mechanics supported his philosophical ideas. He saw quantum entanglement not as a weird exception but as a hint at the fundamental interconnectedness of all things. When particles are entangled, they behave as a single system regardless of distance, which Bohm saw as evidence of an implicate order, which may have been initially inspired and prompted by his previous work with electrons and plasmas.

Subsequently, Bohm extended these ideas to human consciousness and society. Through the rheomode, he developed a form of dialogue where people suspend assumptions and think together, reflecting his view that separation is an illusion.

Consequently, through his work, and during my study of this ever-expanding quantum breakdance, I was led to suspect that neutrinos are the key to unlocking mysterious quantum substrates.

PART II

Consciousness & Cosmic Connections

CHAPTER FIVE

Are Neutrinos the Key
to Consciousness and the Cosmos?

The Big Bang theory is widely accepted by almost everyone on the planet, particularly credible scientists, as the defining event that created all matter and set it into motion. Quite simply, the Big Bang theory states that the universe started from a single point, expanding and continuing to stretch and expand, even now.

Physicists use this theory to discover the fundamental building blocks of the universe and their reactions/interactions. But what holds it all together? What keeps galaxies and humans from spinning out of control and how did an asymmetry take place, a matter-anti-matter *dis-balance* that accounts for manifesting matter?

Uncertainties abound about the early universe. Discrepancies over observations like the cosmic microwave background radiation (horizon problem), a lack of a complete understanding of dark matter and dark energy, and the possibility of alternative explanations for observed phenomena (redshift of galaxies), challenge the idea of an expanding universe from a single pinpoint of origin.

Our difficult puzzle out of the void becomes even more convoluted when we discover many missing pieces to the Big Bang theory, which struggles to completely explain the rapid formation of galaxies in the early universe. Our cosmic speculation still does not explain how gravity controls the universe—and this devotion to abstraction in quantum physics fails to consider the unfathomable concentrated energy found in deep space, one astronomers discovered.

In fact, if anything goes against a gravitational model, we seem to invent or postulate something, including imaginary particles, energy, matter, and waves, to force a fit with gravity. *Why?* For me, this is where the remaining gaslighted shoe falls and shatters.

Consequently, this expansion phase, also known as *inflation,* appears to violate the known laws of physics, including the speed of light limit. But not really. We are told that the only thing that appears to violate the speed of light limit is space. So, again, mainstream science is saved when it uses inflation theory to resolve the horizon problem. Anything traveling through space cannot exceed this limit...unless maybe...the universe exists as electric or electromagnetic in nature rather than depending on gravity. Then physicists tell us that space itself might not be empty at all but consists of a frothy or fluid-like *something.* If that something exists, I suspect it moves faster than the speed of light.

Unfortunately, scientists also lack a method to measure and understand extreme conditions of an early universe, which tells me, we use the most precise tools we have and must still apply our *best guess.* Then we fudge a little by coming up with things we can't prove exist such as *dark matter* and *dark energy,* to name but only two out of many. A few scientists have proposed alternative explanations for how everything we think we know came to be, including, maybe, our conscious awareness.

For example, Plasma cosmology is an alternative model to the Big Bang theory, which suggests plasma and electromagnetic forces, rather than gravity, are the primary drivers of structure formation in the universe.

Even Bohm's "Implicate Order" suggested that the underlying reality of the universe is a dynamic quantum potential having aspects like electromagnetic fields. Quantum potential implies a fundamental role for electrical phenomena at a deeper level of reality, guiding the behavior (voltage, current, and frequency) of particles through a conductor.

Nikola Tesla, like Bohm, believed that everything in the universe is part of an interconnected energy field. By understanding various frequencies and vibrations within this field, we gain deeper insight into the nature of reality. Tesla also emphasized the importance of "resonance," where different systems interact with each other based on their vibrational frequencies, creating a harmonious balance. His ideas, like Bohm's, extended beyond the limitations of classical physics, suggesting that energy could be manipulated and wirelessly transmitted through the air, full of charge, and at no cost, utilizing principles of oscillating energy at varying levels (vibrations).

The dark vacuum-looking spaces between galaxies, planets, stars, and atoms isn't full of nothing at all but teeming with energy, where matter pops in and out of existence all the time, and our universe and the multiverses around it exist in a participant-observer-dependent plasmonic substrate of potentiality.

Cue the music: some unacknowledged covert projects have harnessed what 'pops' into and out of space. And some ancient technologies, like those recently re-discovered at the Pyramids of Giza are but one of quite a few ancient sites all over the world—mostly hiding far below ground, which hold secrets to our humanity—secrets I'm told, that would upend the technological human grid system to which we grow ever more dependent upon, subservient, and built into.

As independent and private as we might believe ourselves to be—we have now come to a point in human existence where it is now only the bravest or poorest among us who possess the courage to live completely *off grid*—without a publicly controlled voltage of current or privacy-probing wave of technology.

Moving on, plasma cosmology theory was primarily developed by Nobel laureate Hannes Alfvén in the mid-20th century. It proposes that most of the visible universe consists of plasma (ionized gas) that conducts electricity and responds to electromagnetic forces. Large-scale cosmic structures like galaxies and galaxy clusters are shaped primarily by electric and magnetic fields rather than gravitational forces and electromagnetic forces, being much stronger than gravity, could explain various astronomical phenomena without requiring dark matter or dark energy.

Subsequently, plasma cosmology *is not* accepted by mainstream scientific communities for several seemingly valid reasons. It fails to explain a lot of observed phenomena that the standard Big Bang model successfully predicts. It can't account for cosmic microwave background radiation or the abundance of light elements. It doesn't provide a comprehensive mathematical framework to match observational data or general relativity—and several of its predictions have been successfully contradicted by modern astronomical observations.

I suppose these predictions *would* fail under a rigid glass slipper of the gravitational model where plasma cosmology will *never* fit. Plasma, like humans, is a complex quantum field system.

While plasma physics plays an important role in astronomical phenomena (like solar flares and cosmic jets), most cosmologists consider the electromagnetic effects to be secondary to gravity in determining the large-scale structure of the universe. But hold on a moment…plasma and neutrinos, while fundamentally different in nature, share intriguing qualities that make them both subjects of interest in highly advanced classified scientific studies.

Plasma is the fourth state of matter, consisting of free ions and electrons. It is electrically conductive and interacts with electromagnetic fields. And it is found in surprising, unexpected places in deep space—I'd even venture to say that plasma is prolific in this universe and beyond.

Alternatively, neutrinos are elementary particles with a very small mass and little, if any, electric charge, which means they rarely interact with matter. Despite this, they pass *through* plasma and can be influenced by its electromagnetic properties under certain conditions. Both plasma and neutrinos are abundant in high-energy astrophysical phenomena, such as stars, supernovae, and black holes.

Plasma is the **DOMINANT** form of visible matter in the universe, while neutrinos are prolific but elusive *carriers* of energy. Maybe this dust mote we call Earth is actually bouncing around or wobbling in *dark plasma*.

Plasma's collective behavior and nonlinear dynamics are now being explored for wave manipulation, as forcefields or shields for stealth aerial, undersea, and floating craft, and for data encoding—which might lend credence for some of the strange UFO-type phenomena people sometimes report. Coincidentally, neutrinos, due to their unique propagation properties, are studied for quantum communication, space travel, nuclear defense, and cryptography to support stealth plasma technologies.

Plasma, due to its unique properties, holds promise in areas such as advanced computing, data storage, and the exploration of consciousness:

I. Plasma for Data Storage

- **High-Density Storage**: Plasma can confine large amounts of data through the manipulation of charged particles. Techniques involving plasma waves and electromagnetic fields allow for information encoding at the quantum level.
- **Durability and Longevity**: Plasma-based data storage could be robust against environmental damage since the information is stored in dynamic, not static, systems.
- **Holographic Data**: Plasma's ability to manipulate light can facilitate 3D holographic data storage systems with unprecedented capacity.

II. Plasma and Information Processing

- **Quantum Computation**: Plasma's nonlinear and chaotic behavior can be harnessed for advanced computation methods, including neural network emulation.
- **Plasmonics**: Surface plasmon polaritons (quasi-particles arising from the interaction of plasma with light) can carry information at nanoscales, offering a pathway to ultra-fast and miniaturized processors.

III. Plasma in Consciousness Studies

- **Brain-Plasma Analogies**: The ionized environment of plasma has parallels to the bioelectrical and ionic activities in neural systems. Some theories posit that consciousness emerges from coherent wave patterns, which can be studied through plasma simulations.
- **Energetic Fields and Thought Models**: Plasma's collective electromagnetic behavior may provide insights into how neural

or bioelectromagnetic fields organize to produce cognition and awareness.

- **Simulation of Complex Systems**: Plasma physics can be used to model complex networks, such as brain activity, offering tools for understanding the dynamics of thought and memory.

IV. Cross-Domain Explorations with Plasma and Neutrinos

- **Information Transfer**: Plasma's electromagnetic properties and neutrinos' ability to penetrate matter could be explored for new methods of information transfer through otherwise inaccessible environments.
- **Cosmic Consciousness**: Some speculative theories explore the interaction of plasma (as a universal medium) and neutrinos (as carriers of universal information) to investigate whether consciousness might have a quantum and/or cosmic basis.

Notably, some scientists admit that modern cosmology is too patchwork for advanced electrical experiments with plasma. Taking a quiet path, their advanced computer simulations show a revolutionary view of the universe. Yet, despite denial and protest, advanced plasma lab research *IS* a thing.

Alternatively, challenges abound when considering plasma and neutrinos for human and technological advancement. Plasma is chaotic and dynamic in nature, and therefore, difficult to precisely control and manipulate, unless...maybe...a neutrino or trillions hold the key, and one has a super high 'need-to-know' secret clearance.

Neutrinos interact weakly with matter, including plasma, making it challenging for the common person to utilize them directly for practical purposes. Furthermore, applying plasma and neutrino physics to consciousness studies requires bridging significant gaps in physics, biology, and philosophy. Would you like some more Kool-Aid?

By harnessing plasma's electromagnetic versatility alongside neutrinos' penetrating characteristics, scientists have uncovered even more revolutionary applications in technology that will deepen our understanding of consciousness and the universe's fundamental information architecture. This is already happening...some clandestine labs, I'm convinced, *can* and do control plasma.

Neutrinos are Second Only to Photons

Somehow, we shouldn't be here. The Big Bang, if there was a Big Bang at all, should have annihilated and obliterated EVERYTHING. But it didn't. Could the smallest measurable particle in the universe be responsible for helping humans arrive and survive on Earth?

Discovered in 1955, neutrinos are nearly massless particles and the smallest particle in our universe, as far as we know. Next to photons (which may be concentrations of neutrinos) they are the most abundant particle in the universe. While *nearly* massless, could a super-abundance of neutrinos dominate the total mass of the universe, despite being infinitesimally tiny?

Moreover, the weak force of neutrinos seems to give them a supernatural property of easily moving through all matter as if it were transparent. All stars, including our sun, produce trillions upon trillions of neutrinos through nuclear decay processes—and since neutrinos rarely interact with matter the way we define interactions, they pass through everything unhindered (by the trillions), including humans.

Consequently, neutrinos exhibit a remarkable behavior known as *neutrino oscillation*, which is the quantum phenomenon of superposition where they change "flavors" (types) as they travel through space. One physicist commented that this oscillation was like "a dog changing into a cat".

This means that studying neutrinos could uncover core secrets about the universe and consciousness. Anytime scientific descriptions are bestowed with magical properties, such as shapeshifting animals being used to describe changing neutrino flavors, my antennae go up. Why? Because the sun's output of neutrinos, a byproduct of nuclear fusion, we're told, is much less than expected using the Standard Model of physics.

It appears more likely that the sun's surface spawns all the flavors of neutrinos, negating the need to account for changing flavors—which is merely another ruse to cover the shortfall. When I pair this up against Russian physicist (Nikoli Kozyrev, torsion fields, nature of time, and rotating systems) experiments that showed natural clock-wise rotating systems adding energy while counter-clockwise rotations releasing energy, I become super excited. Imagine the security protocols that might be implemented using counter-clockwise systems!

"Foolishness!" I'm told.

Intriguingly, neutrinos upend and are likely responsible for why we, and everything else in the universe, exists. And neutrinos cannot, we're told, be contained. Since they rarely interact with anything, they are considered *weak* due to a mass of around 0.8 electron volts (eV), which is significantly lighter than an electron. But even this is inaccurate because a neutrino's exact mass is not known.

However, we're getting closer.

In April 2025, researchers at the University of North Carolina, Chapel Hill, revealed their most precise measurement yet of a neutrino in the journal *Science*, scaling down its mass. The value was found to be no more than 0.45 electronvolts—a million times lighter than an electron. If we nail down the mass of one type of neutrino, the other flavors could be calculated, leading to a new physics beyond the Standard Model.

And what if this sorcerer's particle is simultaneously both matter and antimatter or an entanglement thereof? This might mean that neutrino pairs form photons—or are photons—and that plasma could be a sort of gathering of neutrino masses.

Neutrinos allegedly come in *flavors,* as previously mentioned: electron neutrino (νe) muon neutrino (νμ) tau neutrino (ντ) and maybe something called a *sterile* neutrino. A sterile neutrino is hypothetical too and might only interact with other particles through gravity, unlike the known "active" neutrinos which interact via the weak force.

The funny thing about one flavor among this entourage, the muon (like an electron but heavier), is that when it was shot through an intense magnetic field at the Fermilab a few years ago, it disobeyed known laws of physics. Basically, the muon wobbled more than expected in the magnetic field.

This muon wobble went against scientific predictions where the measured magnetic moment of the muon deviated significantly from the prediction of the Standard Model of particle physics. What does this mean? It means that muons are likely interacting with unknown particles or forces not accounted for in our current and public understanding of physics.

"Not so fast!" Physicists say.

While the Fermilab results mirror those of a Brookhaven National laboratory experiment back in 2001, this Fermilab experiment fell short of the scientific gold standard to be declared new physics. And lacking the funding to continue, Brookhaven allegedly retired its fifty-foot muon storage ring post experiment, to the disappointment and suspicion of truth crusaders everywhere.

Why would a lab like Brookhaven swiftly retire such an important piece of equipment and shut down replication efforts on such a game-changing finding? It didn't. It just moved the experiment deeply underground.

What's more, while muon experiments continue, it's what's *not* being said that I find most interesting. While the experiments have, fortunately, continued in other labs, they've gotten more precise. Yet the reports appear vaguer. The BLUF (bottom line up front) is that they don't tell us much outside of increased experimental precision. Did muons wobble the Department of Energy's clandestine comfort zone a bit too much? That's another book.

Thankfully, the *Muon g-2 Theory Initiative*, consisting of approximately one-hundred seventy experts with deep knowledge of standard and quantum physics, concluded after three years of intensive workshops and calculations using the Standard Model, that the peculiarities of the Brookhaven muon anomaly within the experiment violated known laws of physics.

Suddenly, an alternative group quickly arrived on scene using a different technique, known as a lattice calculation. This new group computed the muon's magnetic moment too and quickly distributed an entirely different conclusion. Zoltan Fodor of Pennsylvania State University, one of the authors of a report published in *Nature,* claimed no discrepancies between the Brookhaven result and the Standard Model, meaning, **NO NEW PHYSICS!** This was the quantum physics version of Project Grudge, without the UFO's.

Muon Vindication

The great news is that researchers performed a new blind experiment in 2022, where they had to relinquish their biases and measurements, because technicians set the muon clock at Fermilab to an unknown rate and locked it away. The results confirmed that the Brookhaven anomaly was not a fluke at all. Joe and Jane Public now have an opportunity to crack the Standard Model of physics! I get shivers just thinking about this.

Moreover, scientists have indications of a matter-antimatter asymmetry of the universe. And many theorists now believe that studying such properties in the neutrino sector could be important for understanding one of the great cosmological mysteries: Why do we, or anything, exist?

Recently, experiments in Japan discovered a telltale anomaly in the behavior of neutrinos, and the results suggest that, amid the throes of creation and annihilation in the first moments of the universe, these particles, while tiny, might have collectively tipped the balance between matter and antimatter. As a result, a universe that started out with a clean balance sheet — equal amounts of matter and antimatter — wound up with an unbalanced excess of matter: stars, black holes, oceans and us. Fair warning: More research is required.

Neutrinos entered the world stage in 1930, when Wolfgang Pauli, a theorist, postulated their existence to explain the small amount of energy that goes missing when radioactive decays spit out electrons. Later, Enrico Fermi, the Italian physicist, gave them their name, "little neutral one," referring to their lack of an electrical charge. In 1955 Dr. Frederick Reines discovered them emanating from a nuclear reactor.

Alternatively, if we could view a neutrino directly moving toward us, I *imagine* it moving in a vortex-like circle and its force equal to its diameter. Some theoretical models propose that subatomic particles exhibit torsion, spin, or helical field behaviors in their motion, especially when moving through spacetime. While neutrinos are known to have spin (they're fermions with a spin of one-half), this "vortex-like" image symbolizes for me how neutrinos tunnel or twist their way through dimensions.

A man named Robert Beckwith thought the same thing—and so did Nikola Tesla, albeit the latter man never referred to them as neutrinos.

Beckwith was an electrical engineer who allegedly worked on a part of the Philadelphia Experiment (a vehemently discredited U.S. Navy experiment that made a destroyer invisible and transported it from Philadelphia to Norfolk in 1943). He was an honored scientist, inventor, researcher and consultant. He had multiple top-secret clearances, with over fifty years as a world leader in the electrical power industry. In 1967, he founded Beckwith Electric Co., Inc., in Largo, Florida. The man invented something extraordinary that we still use today: The national power grid.

Beckwith Energy was acquired by Hubbell Utility Solutions in 2020, making it part of the larger Hubbell Incorporated. The recent acquisition of Systems Control by Hubbell for $1.1 billion underscores Hubbell's strategic growth initiatives and Beckwith's quantum business prowess.

Anyway, Beckwith, like da Vinci, Telsa, and Bohm appeared quite comfortable hanging out on the fringes of science. Beckwith practiced trying to levitate quarters, drove metal rods into the ground to *hear* earthquakes in the Pacific Ocean, and believed neutrinos to be the voice and power of the universe. He once wrote a paper with a very hard-to-locate man named Daniel (or Drew) Craig, entitled *Hypothesis*.

Hypothesis is a treasure trove of information regarding super atoms, ball lightning, neutrino-based telepathic communication, divided space, levitation, advance weapon systems, and a host of other topics including crop circles and body implants. He very much believed in extraterrestrials, which he considered intelligent inter-dimensional beings and claimed humans had the ability to "see" or read while blindfolded and could levitate or time-travel.

In fact, his patent, *Neutrino Light to Photon Light Converting Matrix* (No. 6,981,310 B2 May 10, 2005) creates pixels for viewing underground structures from a satellite. In the patent he states, "We do not believe the Big Bang Theory is valid." He postulated two fields of force lines emanating from the nucleus of all atoms, *near and far force*, where near lines repel other atoms to prevent destruction. The patent goes on to apply engineering to the neutrino.

Neutrinos are like tiny neutral vortexes capable of interdimensional-interstellar travel, not superpositioned as in a dual existence, but always arriving and leaving as a sort of information-gathering *wave packet*. This is in line with Louis de Broglio's assertion that neutrinos are toroidal vortex condensations of the medium of an electromagnetic field. Broglio believed that photons consist of two neutrinos and two anti-neutrinos—and any attraction between neutrinos within photons and electrons is a vortex interaction known as the Bernoulli Effect.

Remember, an atom, which is HUGE compared to a neutrino, has several strong force lines emanating from its nucleus. Some of these force lines hold orbiting electrons or link with other atoms to form molecules. Some molecules further link up, forming solids, liquids, gases, or plasma.

Conversely, other particles allegedly extend across the universe holding it together. A neutrino is, in this simple fiction, a vortex of weak force, but not energy. And if we could stop a neutrino in its tracks, its mass and energy would be zero, because it would cease to exist. It's like trying to hug a ghost.

As neutrinos pass through the universe, they are *jerked* by the weak forces of atoms as they pick up information concerning atoms. What is thrilling to consider is that the action of DNA molecules inside sentient beings produces nearly the same information as neutrinos going *through* our bodies. Then, information, carried by neutrinos becomes scattered in all directions, thus making sentient beings' receivers, containers, and carriers of information.

Another exciting experiment, while not exactly related to neutrinos shows that humans have successfully made atomic particles greatly expand and *spin* (aka *angular momentum,* a description for something that moves without traveling anywhere) on command!

Dusan Sarenac's research team at the University of Waterloo in Ontario, Canada realized this breakthrough, which was published in the journal Scientific Advances (*Experimental Realization of Neutron Helical Waves 18 November 2022*). It is touted as a potential leading factor in the development of quantum supercomputers.

Sarenac works with neutrons, located in the center of atoms. Neutrons are electrically 'neutral' and can pass through materials that block light and X-rays. Notably, and to understand scale, we can lay more than ten-thousand neutrons across a single strand of human hair.

Here's where it gets interesting…another experiment by physicists now suggests, after measuring the radioactive decay of a substance known as beryllium (which later turned into lithium), that neutrinos might be considerably larger than an atomic nucleus! (*Beryllium Electron capture in Superconducting Tunnel junctions' experiment (BeEST) 2025*)

The results of this latter experiment, performed by an international group of physicists are published in the journal *Nature.* The findings suggest that the neutrino's *wave packet* is localized at scale significantly larger than a typical atomic nucleus, offering major insight into the quantum properties of neutrinos.

Both experiments push the boundaries of understanding matter at quantum and subatomic scales, revealing new insights into particle behavior and interactions.

In the neutron experiment, we learn how neutral particles interact with matter, while the neutrino experiment takes this a step further by investigating the scale and properties of neutrinos, whose wave packets appear surprisingly large for such tiny particles. But these experiments, although seemingly advanced, pale in comparison to classified technologies developed, built and implemented decades ago by first world partnerships between governmental and private sector entities.

Let's take a quick peek.

Neutrino Storage Capability

Do neutrinos have a memory storage capability?

Neutrinos are in a frequency range of 10 to the 33rd Hz.

The information carrying capability of neutrino light appear to be decimal orders of magnitude greater than photons! Photons can and do encode information in their quantum states, such as polarization, phase, and frequency. These states are used to represent bits of information (0s and 1s in binary) or even qubits in quantum computing, enabling vast amounts of data to be processed simultaneously.

Photons are already used in fiber-optic communication, where they transmit information over long distances with minimal energy loss. This makes them highly efficient for communication. Additionally, in controlled environments, photons maintain their quantum state for extended periods, which is critical for certain forms of quantum memory.

Photons travel at the speed of light (in a vacuum), making them incredibly efficient carriers of information. Remarkably, Chinese scientists recently measured a pulse of light in *thirty-seven dimensions*—a stunning testament to the complexity, (nonlocality), and potential of light.

Perhaps, in a metaphorical sense, we too exist in a sort of vacuum— be it scientific, spiritual, or otherwise. This might partially explain our struggle to uncover definitive quantum truths in a space that appears empty but isn't. Proof of full space arrives in calculations of zero-point energy, which suggest that just one cubic centimeter of "empty" space contains more energy than all the matter in the known universe! So, is it truly empty in the sense we think of as consisting of nothing?

Indeed, there exist a few legitimate issues with harnessing photons for memory storage or information processing. Photons must interact with atoms to encode or store data for extended periods. Furthermore, photons are notorious for how easily they scatter, and the risk of information being carried by them and getting lost is super high. As we learned in the coherence chapter, photons, like neutrinos are master escape artists capable of decoherence. Each can *poof* and disappear. They require extreme isolation and precision.

While photons are incredibly adept at transmitting information and hold promise for advanced memory systems (especially in quantum computing), they are not *naturally* suited to *store* memory in the traditional sense.

However, clandestine advancements in quantum and optical technologies have overcome many of these limitations, making photon-based memory systems viable and in use but not publicly available—yet—at least until something better is discovered and built.

Now, some people out there in high (or low) places will publicly scoff at my "systematic analysis" of accidental exposures as the stuff of conspiracy theory. But as I stated in the beginning, I write fiction (and consider myself more of a curiosity crusader)—so these realizations shouldn't raise too many alarms if untrue, meriting only a shrug of indifference, or like Bohm and Tesla, going completely ignored.

Moving along, if the Big Bang really did occur, how many neutrinos existed prior to our universe? And how many escaped the Big Bang?

If what I'm pondering, as I watch ripples from a pond in front of me, is true, the neutrino field would include all information about the universe, like an Akashic field, a nearly infinite library of information contained in some sort of harmonic resonance code. And if the universe is expanding, the way we are told it is expanding—do neutrinos multiply and expand too? If they do, how?

The answer we're told, is no, neutrinos do not expand with the universe. However, we do not have a direct way to test whether neutrinos themselves expand with the universe in the same way that spacetime does (although I have my doubts about spacetime expanding too). The number of neutrinos, we've been told, remains *roughly* constant, and their density decreases as the universe expands. Neutrinos permeate space as cosmic neutrino background like cosmic microwave background radiation.

I beg to differ.

Plasmas and explosions in the universe indeed give birth to more neutrinos, particularly under specific astrophysical conditions. In plasmas, high-energy particles (like protons, electrons, and atomic nuclei) collide at extreme energies where such collisions lead to nuclear reactions, particle decays, or secondary processes that produce neutrinos.

For example, proton-proton fusion in stars produces neutrinos as a byproduct (e.g., in the Sun, this accounts for solar neutrinos). Also, in extremely hot plasmas, photons interact with electrons or nuclei, leading to pair production or decays that generate neutrinos. In a supernova explosion (e.g., a core-collapse supernova), an immense amount of energy is released, including in the form of neutrinos.

During the collapse of a star's core, protons and electrons combine under immense gravitational pressure to form neutrons and neutrinos. Approximately 99% of the energy in a supernova is carried away by neutrinos! These neutrinos are emitted in enormous numbers and escape stars relatively easily, unlike photons or heavier particles.

Likewise, when cosmic rays (high-energy protons and nuclei) interact with interstellar matter or radiation, they produce pions (π particles of the meson family that are quark-anti-quark particles) that decay into neutrinos. Such processes typically produce large fluxes of high-energy neutrinos, particularly in astrophysical environments like gamma-ray bursts (GRBs), active galactic nuclei (AGN), or shocks in interstellar space.

Similarly, events like black hole mergers or neutron star collisions involve extreme energies, where nuclear matter is heated and compressed, leading to neutrino production. In such an environment, neutrinos serve as core messengers that carry information about the underlying processes.

So, why were we told that the number of neutrinos in the universe remains relatively constant in an expanding universe and exist primarily as cosmic *background* fodder? Because it keeps the proverbial playing field of Alex Wendt's theory of International Politics quite level (see Intro) and public scientists won't stray too far from their federally funded padded labs and playpens.

Interestingly, a "neutrino energy power cube", designed by Holger Thorsten Schubart, the founder and CEO of the Neutrino Energy Group, is headed, the company claims, for eventual mass production.

The cube is a device designed to generate electricity by capturing a small amount of kinetic energy from naturally occurring neutrinos, using a technology called neutrinovoltaics.

The product aims to produce a continuous, weather-independent power source by harnessing energy from these particles. It is a compact unit containing specialized materials like vibrating graphene lattices to interact with the neutrinos and convert their energy into electricity. If Schubart and his cohorts scientifically succeed, it will be a sacred miracle of exponential proportions.

Tesla's Aether (Ether)

Nikola Tesla, a scientist, engineer, visionary, and inventor in the 1900's, believed in the existence of "ether" (or "aether") as a medium for electromagnetic waves and what he called "primary solar rays". Neutrinos weren't discovered until after his death—but it's likely he was onto them.

As previously mentioned, the sun produces neutrinos primarily through the proton-proton chain reaction, which converts hydrogen into helium in its core. This process generates electron neutrinos.

And there exist intriguing parallels between Tesla's ether and neutrinos: Both were proposed as invisible, pervasive substances/particles that fill space. Tesla's ether, like neutrinos, was conceived as an "all-pervasive" medium. Likewise, neutrinos literally pass through everything. And each, ether and neutrinos are connected to ideas about free energy transmission.

To boot, Tesla saw ether as absolutely essential for the wireless transmission of energy, in limitless supply, and readily available for harvesting. It is interesting to note that Einstein altogether omitted *aether* in his theory of relativity because, sensibly, it wouldn't fit into a gravitational model—but it lands softly and upright on the dock of an electrical model of the universe. It's also notable to remember, that scientists at the time, considered Tesla a fool (or a serious patent rival) and scoffed at his idea of global wireless energy. It was the beginning of the end for Tesla and his career, but he knew...he knew.

Moreover, Tesla viewed Earth not as a planet, but as a "system environment" and likened it to a Tesla coil. The sun and moon, he said, are wirelessly powered through Earth's electromagnetic field and the only force needing to be countered is electromagnetic rather than gravitational. Would this render space more like a lattice of photon filaments held together by a fluid of plasmic material or maybe neutrinos rather than a warped fabric caught in time?

Tesla also had a fascination with the numbers three, six, and nine. He believed that these numbers were a blueprint for the laws of energy and motion.

What few people know is that Tesla and Dr. Heinrich Hertz engaged in heated debate regarding the rigidity of space and waves. Hertz claimed that a medium of structureless ether filled all of space and he presented polished mathematical results held up by the world as gospel. Conversely, Tesla proved that space is filled with waves propagated by alternating compression and expansion, which to me, sounds a lot like Bohm's later idea of folding and enfolding.

Tesla's experiments contradicted Hertz's elegant math because they definitively showed that Tesla's oscillators could produce more efficient frequencies for wireless transmission. Yet, to this day, wireless frequencies are known as Hertzian waves, but they really belong to Tesla.

Notably, the 369hz frequency is also considered a blessing in Hinduism (Oppenheimer's adopted religion). It is believed to resonate with the third-eye and crown chakras, which are crucial for higher consciousness and enlightenment. Many Hindu temples are constructed with numerical symmetry in mind, reflecting cosmic harmony. Even The sound **"Om"** (**Aum**) is considered the primordial vibration of the universe that stimulates the vagus nerve in humans.

Some modern spiritualists connect frequencies like 369 Hz to resonance and harmonic properties, which enhance healing and emphasize the importance of vibrations in spiritual awakening. It is said the frequency synchronizes brain waves.

Comparatively, scientist Marko Rodin postulated that the numbers three, six, nine represent a "flux field" or a vector from the third to the fourth dimension. Furthermore, Tesla experimented with what he called "longitudinal electric waves" or "scalar waves". He thought these waves might be a form of electromagnetic radiation, where the oscillations occurred in the direction of propagation, rather than perpendicular like traditional transverse electromagnetic waves.

Tesla believed these waves could travel faster than light, penetrate solid objects more effectively than traditional EM waves, be used for wireless power transmission, and carry information over long distances—like neutrinos.

J.P. Morgan, an oil baron making shitloads of money on a variety of global investments, pulled Tesla's funding when Tesla excitedly shared with him that energy could be free for everyone on the planet. I suspect Morgan, rather than laughing at Tesla, was horrified and quickly realized the detrimental implications free energy would have on Morgan's family, friends, future and fortune.

Additionally, Thomas A. Edison, Tesla's electrical rival was much more of a businessman and public relations guru than Tesla. While Tesla is credited for the A/C current we use today, Edison was likely still reeling over his loss to have D/C become the standard for electrical power—thanks in part to Tesla and George Westinghouse—and worked behind the scenes to further discredit Tesla's visionary free energy ideas.

Since Morgan and his cronies risked having their wealth upended and their pedestals toppled if electrical fields were leveled and readily available at no cost to common folk, something had to be done about Tesla. As an elite with incredible power and influence, Morgan easily left Tesla and the rest of us in the dark—unless we paid to see the light. And here we remain to this day.

No matter what device we plebians use, at some point, it must be plugged in to some outlet and recharged, and someone always pays the price to an electric company or a government for what could, and perhaps, should be free. Land, food, water, and healthcare fall under the same rigidly controlled purview. Each, and the parties which manage them, is a complex system nearly impossible to saddle and control for greater good. And sadly, most of it is wasted in favor of 'giving it away'.

Tesla died in relative obscurity, and it is only recently that his patents, theories, and ideas are being resurrected for further study. Unfortunately, the most important aspects of his research were quickly seized, classified, and eventually sent to Wright-Patterson Airforce Base in Dayton Ohio. Here, under the command of Brigadier General L.C. Craigie, in 1945, *Project Nick* was heavily funded to test Tesla's research on particle beam weaponry, among other plasma research—namely, controlling the latter.

The results of these experiments, and others, have never been published. It is said by government officials that Tesla's seized research was utterly worthless, and that he was a delusional quack. Then, why confiscate any of Tesla's documents in the first place? Or why not declassify the results and papers from *Project Nick*?

Comparatively, the high-frequency, short duration and electrical discharges associated with Tesla coils are known to create high strangeness with penetrating non-X-ray radiation dubbed Exotic Vacuum Objects (EVO's) by researcher Ken Shoulders. Shoulders discovered that EVO's can encapsulate and transport matter Low Energy Nuclear Reactions (fission).

In fact, and according to an article in Newsweek, a U.S. warship deployed in the Western Pacific Ocean, fired a laser at a drone target in February of 2025. The *Preble* is but one U.S. destroyer armed with a high-energy laser weapon ala Nikola Tesla-style. A laser uses energy fired at the speed of light, which is likely less expensive per shot and has virtually unlimited firing power compared to traditional defensive ship-based weapons like missiles. While this *HELIOS* of a laser can take down simple drones and craft, imagine what a scaled up and manned satellite laser orbiting Earth could do!

As a side note, I learned while writing this book that about seventy percent of *all* government documents are classified *just because* (but it's really to prevent something called *accidental exposures)* and much material remains so ad infinitum—the stuff we pay for with taxpayer dollars that could, maybe, kill us all or keep us healthy and alive way past the age of one-hundred-twenty. Keep working.

Returning to more interesting matters, both ether and neutrinos were initially very difficult to detect, and impossible to prove during Tesla's time, leading Einstein to reject *ether* in his theory of special relativity. But Tesla's ether didn't fit into a gravitational model because much of Tesla's work was grounded in the study of electricity, electrical waves, oscillation, and plasma.

Neutrino Detectors

No book about weird quantum processes and elusive, spinning, expanding subatomic particles would be complete without mentioning neutrino detectors. Neutrino detectors are massive scientific instruments designed to detect neutrinos, and while conceptually serving a completely different purpose than Tesla's wireless energy towers, we build them:

1. *To understand fundamental physics and the nature of the universe*
2. *To study nuclear processes inside stars and supernovae*
3. *To monitor nuclear reactors and non-proliferation*
4. *To test theories about dark matter and other cosmic phenomena*
5. *To study Earth's interior (geoneutrinos to study Earth's crust and mantle)*
6. *To search for new physics*

These detectors must be constructed deep underground or in the ocean for crucial shielding against decoherence. The detectors need to be protected from cosmic rays and other background radiation that would interfere with neutrino detection. A mile of rock or water can filter out most unwanted particles while allowing neutrinos to pass through unimpeded. This is why you'll find these detectors in mines, under mountains, or in the deep ocean.

However, there exists a much simpler and far less costly and disruptive way to capture neutrinos. And we are receiving obfuscated conjecture regarding their true purpose—but that too, is beyond the scope of this book.

Major neutrino detectors include:

1. **Super-Kamiokande (Japan)**
   - Located 1000 meters underground in the Kamioka mine
   - Uses 50,000 tons of ultra-pure water
   - Notable for helping to prove neutrinos have mass

2. **Daya Bay Reactor Neutrino Experiment** is a multinational particle physics project in China studying neutrino oscillations. This collaboration includes researchers from China, Chile, the United States, Taiwan, Russia, and the Czech Republic. A portion of the project is funded by the US Dept. of Energy's Office of High Energy Physics.

3. **IceCube (South Pole)**
   - Built deep in Antarctic ice
   - Considered the largest neutrino detector, it uses a cubic kilometer of Antarctic ice to detect neutrinos from cosmic sources
   - IceCube Gen2 is expected to come online and is ten times larger than IceCube
   - Recently, in a search for extremely high energy (EHE) neutrinos, researchers spotted events where horizontally travelling neutrinos deposited a huge amount of light inside the detector. This reduced the background noise of atmospheric muons. What Maximilian Meier and Brian Clark discovered is that cosmic ray flux is mostly composed of elements heavier than protons. This research offers answers to something scientists have been trying to answer for more than one-hundred years

4. **SNO (Sudbury Neutrino Observatory, Canada)**
   - Located 2km underground in a nickel mine
   - Used heavy water as detection medium
   - Solved the "solar neutrino problem"

5. **DUNE**
   - South Dakota
   - Will use liquid argon technology
   - Designed to study neutrino oscillations in detail

6. **KM3NeT, Cubic Kilometer Neutrino Telescope (Mediterranean Sea)**
   - Network of light collecting glass sphere detectors built under the Mediterranean
   - Will use seawater as detection medium, instead of hyper-pure or heavy water used by others
   - Aims to study both cosmic neutrinos and neutrino properties
   - Recently, astrophysicists observed the most energetic neutrinos (muons) ever, carrying *unprecedented* energies of an astounding 120 PeV's

7. **The Jiagmen Underground Neutrino Observatory**
   - Situated 700 meters underground in Jiangmen, Guangdong province, China
   - A large acrylic sphere, considered the project's centerpiece, is the world's largest transparent spherical detector designed to capture neutrinos

8. **The "Radio Neutrino Observatory Greenland"** (RNO-G), which uses the Greenland ice sheet to detect ultra-high energy neutrinos by detecting radio waves emitted when neutrinos interact with the ice

9. **The Irvine-Michigan-Brookhaven Detector**, deep underground in a salt mine outside of Painesville, Ohio, picked up a supernova explosion spraying neutrinos through space, kickstarting the field of neutrino detection after researchers witnessed eight neutrinos within five seconds. It is the framework upon which most later neutrino detectors were built

10. **MiniBooNE**, BooNE stands for Booster Neutrino Experiment and is part of DOE's Fermi National Accelerator Laboratory. This detector picked up an excess of electron

neutrinos in 2006 that conflicted with US Department of Energy's Los Alamos National Laboratory (LSND) results but still indicated the possibility of sterile neutrinos. (Later runs were more consistent with LSND.)

11. **Los Alamos' Liquid Scintillator Neutrino Detector** picked up excess neutrino oscillations where theory predicted there should be none in 1995.

12. **Three I Can't Mention**

Neutrino Detectors seek to uncover:

- The precise mass of neutrino *flavors* (muon, electron, tau or sterile)
- Whether neutrinos and antineutrinos behave differently (*Charge Parity* violation)
- The role of neutrinos in cosmic events like supernovae
- Potential signs of dark matter
- Whether neutrinos are their own antiparticles (they are)
- Sources of high-energy cosmic neutrinos
- Details about the nuclear fusion processes in the Sun

The detection methods vary but often involve massive volumes of highly purified water, ice, or liquid argon surrounded by extremely sensitive light sensors. When a neutrino finally interacts with an atom in these materials, it creates a flash of light (a photon) or other detectable signal that scientists then analyze.

These detectors are among the largest and most costly scientific instruments ever built but—there is an easier and far less expensive way to capture and use neutrinos, one Shubart is close to cracking for commercial use.

The Weak Neutrino

Again, a **neutrino** (*v*) is an elementary particle that interacts via weak interaction and gravity. A neutrino's handedness is always *fixed*. As I mentioned in the previous chapter, regarding asymmetry, this violates parity and only left-handed chiral states (and right-handed antineutrinos) participate in weak interactions according to the Standard Model of particle physics. But remember, that model's been cracked thanks to what happened at Brookhaven.

So, the neutrino is electrically neutral because its rest mass is so small that it was long thought to be zero. The rest mass of the neutrino is much smaller than that of the other known elementary particles (excluding massless particles). The weak force has a very short range, the gravitational interaction is extremely weak due to the very small mass of the neutrino, and neutrinos do not participate in electromagnetic or strong interactions.

Thus, neutrinos typically pass through normal matter unimpeded and undetected. In fact, billions of them are passing through your fingernails right at this moment! Due to this, neutrinos are usually considered void prospects for propelling consciousness, UFOs, or much other information. We are told that life and energy take cues from largely biochemical and/or electromagnetic processes and neutrinos have little to do with it. This is what *they* want you to believe. It helps keep the top-secret, *eyes-only* yellow brick road paved with gold.

Neutrinos absolutely carry energy. In fact, neutrinos are produced with varying energies through different natural and artificial processes:

1. Solar neutrinos from nuclear fusion in the Sun typically have energies in the MeV (million electron volt) range

2. Atmospheric neutrinos, created when cosmic rays hit Earth's atmosphere, can have energies up to TeV (trillion electron volt) range

3. The highest-energy neutrinos ever detected came from cosmic sources and were observed by the IceCube Neutrino Observatory in Antarctica, with energies in the PeV (petaelectron volt) range—that's a thousand trillion electron volts!

   A) If we could somehow capture and store this energy, the amount would be significant.

   B) Theoretically, harnessed neutrino energy would provide enormous thrust for spacecraft

   C) Since neutrinos pass through most matter, they could, again theoretically, deliver energy to very specific locations

   D) Neutrinos could escape from the core of supernovas before light does (think about that for a moment and pair it up against Einstein's Theory of Relativity).

This is outstanding to learn, especially because neutrinos interact so weakly with matter. Remember, neutrinos are phantom-like and difficult to detect. We might view them as the holy spirit of Source moving through EVERYTHING. However, they gain or lose energy through various interactions, which can be remembered through the acronym **OSI**.

Oscillation happens between different neutrino flavors (electron, muon, and tau neutrinos), meaning they change flavors (but maybe not), *Scatter* off other particles and *Interact* with strong gravitational fields.

The study of high-energy neutrinos is particularly important in astrophysics, as they carry information about some of the most energetic events in the universe, like supernovas and active galactic nuclei.

Regarding space travel, neutrinos could potentially enable:

1. **Navigation:** Since neutrinos travel in straight lines from their sources (like stars) and pass through most matter, they could theoretically provide an extremely precise navigation system for deep space travel. Unlike conventional systems that rely on visible light or radio waves, neutrino-based navigation wouldn't be affected by dust, radiation, or other interference.

2. **Communication**: Their ability to pass through solid matter means neutrinos could potentially enable communication through planets or other obstacles. This could solve the current problem of losing contact with spacecraft when they pass behind celestial bodies or the monitoring and detection of deep ocean nuclear submarines or nuclear activity in deeply buried underground reactors.

3. **Nuclear Monitoring**: Neutrinos could be used to detect nuclear submarines where they shouldn't be or identify and pinpoint nuclear testing or hidden reactor sites.

(Non?) Delusions of Neutrino Grandeur

For the da Vinci's and tomb raiders among us, let's engage in a lofty Tesla-Crisp *dream-time* experiment on the even greater hypothetical applications of neutrinos, for humans and others, who capture and store them:

1. **Deep Space Power Generation**
   - Since neutrinos pass through virtually everything and come from all cosmic sources, spacecraft could generate power anywhere in space, even in the darkness between stars
   - This would revolutionize deep space exploration by removing reliance on solar power or nuclear generators

- Could and has enabled indefinite-duration missions to the outer solar system and beyond

2. **Stealth Energy Infrastructure**
    - Since neutrinos pass through Earth effortlessly, we could build power plants deep underground
    - Cities could be powered by facilities kilometers beneath them, freeing up valuable surface space
    - Underground power networks could be completely hidden and protected from weather/disasters

3. **Communication Technologies**
    - Since neutrinos pass through planets, we could achieve instantaneous communication through Earth
    - Submarines could communicate while deeply submerged
    - Regarding neutrino applications, the technology could enable indefinite-duration missions to the outer solar system and beyond.

4. **Solar Core Monitoring**
    - By capturing and analyzing solar neutrinos, we could get real-time data about fusion processes in the Sun's core
    - This could give unprecedented warning of solar events and better understanding of stellar physics

5. **Novel Propulsion**
    - If we could not just capture but also direct neutrino energy, it could enable new forms of spacecraft propulsion
    - Since neutrinos carry momentum despite having nearly zero mass, they potentially hold promise for reactionless drives

6. **Geological Exploration**
    - By analyzing neutrinos passing through Earth, we could effectively "scan" the planet's interior

- This could revolutionize mineral exploration, volcano monitoring, and earthquake prediction

7. **Portable Power**
   - Personal devices could potentially run forever on ambient neutrino energy and there would be no need to charge phones or other electronics as they'd constantly harvest neutrino energy

8. **Medical Applications**
   - Highly targeted energy delivery through tissue could enable new forms of medical treatment
   - Neutrinos might be able to power internal medical devices without batteries

9. **Time Capsules**
   - Since neutrinos interact so rarely with matter, we might store information in neutrino states
   - Neutrino-state storage would create nearly eternal information systems

10. **Fusion Power Enhancement**
    - Captured neutrino energy could be fed back into fusion reactors to improve efficiency
    - Neutrino capture would solve one of the major energy loss mechanisms in current fusion reactor designs

Any one of these Tesla-Crisp *dream-time* scenarios would fundamentally change our relationship with energy—making it universally available and essentially limitless. It would be comparable to having a perpetual motion machine (though not violating physics since we're harvesting existing energy rather than creating it). Viva la Tesla!

Of course, this is all highly speculative truth crusading under open-domain physics and mathematics. It would require physics-breaking technology and classified mathematics to pluck energy out of air or space, right?

Again, for the da Vinci's, Tesla's, and Bohm's among us, who hope I address such alternate dimensions in the public realm, all of this is more science fiction than scientific. But for the sake of science fiction and imagination, let's explore a couple of possibilities.

Intriguingly, perhaps neutrinos are quantum messengers from parallel universes. Because they barely interact with our world and phase-shift through matter and energy with ease, neutrinos could be entangled with "shadow matter"—particles from a hidden sector of physics, possibly linked to other dimensions. If this were true, then neutrino oscillations likely aren't switching flavors—but leaking information between realities—nature's unseen and advanced communication system between universes—and our detecting said would prove their existence.

CHAPTER SIX

UFO's and other Phenomena

Unidentified aerial phenomena (or whatever we're now calling such things), seem, according to witnesses, to defy our current understanding of classical physics. They pop in an out of existence and commit abrupt maneuvers that would render any humans inside about as smashed and juicy as a bug on a windshield. And they sometimes appear plasma-like, able to shape-shift. Let's imagine that *super atoms* could be designed that control both magnetic lift and time.

Indeed, years ago, a team of physicists led by Eric Cornell at the National Institute of Standards and Technology, cooled rubidium atoms to such a reduced temperature, that the atoms assumed a new quantum state, forming a sort of *super atom*. Rubidium is a highly reactive alkali metal with unique quantum properties. It has potential roles in both magnetic levitation (lift) and time-related phenomena, particularly in the context of quantum mechanics and atomic physics.

Moreover, rubidium's properties could be relevant in creating or stabilizing time-travel devices. Its behavior in ultra-cold states and under electromagnetic manipulation could intersect with theories involving spacetime curvature or wormholes.

Thinking further, control voltages of super atoms might only require minute amounts of energy to power a space vehicle weighing tons, its vibration causing such a vehicle to merely hum rather than roar. Planes and ships weigh tons too, but their weights are bolstered by buoyancy and propulsion—the stuff of classical mechanics. Might UFOs be the stuff of electromagnetic lift—like humans (proverbially and consciously speaking)?

Sometimes I feel that UFO's occasionally present as deceptive holograms or plasma drones—decoys to throw the public into a tailspin that can later be written off as *mass hysteria,* swamp gas, or mental illness. Remember, what the government or its intelligence agencies don't say speaks much louder than its psyops, misinformation, lack of disclosure, heavily redacted declassified documents, and politics.

Pushing ahead, would one frequency, say, a 7.5 Hz signal produce a very calm 'inner' space of our imagined space vehicle? A 7.5 frequency basically bounces a transmission around the world right back to its starting place and it produces theta waves, also useful for meditating, daydreaming, and hypnosis. This might explain why some UFO experiencers feel locked in place, unable to move, or ambling toward such objects when they appear, without fear.

What if we used a higher frequency and lower amplitude causing the surface of any sort of *otherworldly* craft to emit light? This sort of energy from the light (perhaps red-orange?) might come from neutrinos causing the energy of electrons to oscillate. Ah, but no, I'm told. Secret science does not allow plebian civilians to play with room-temperature super atoms so they may advance crackpot theories about dimensional plasma-like spacecraft. So, does that mean we *have* room-temperature super atoms? Could, say, neutron thrusters enable faster than light-speed travel? More on that later...

What about Brownian-motion propelled craft?

When gas molecules collide with a surface, they typically bounce off with random velocities following something called the Maxwell-Boltzmann distribution. However, if we could design a special surface that interacts asymmetrically with gas molecules—one that reflects molecules differently based on their approach angle or velocity—we could create a net force in one direction.

This is conceptually like a Brownian ratchet, where random molecular motion is rectified to produce directed motion.

Consequently, Brownian motion refers to the random, jittery movement of microscopic particles suspended in a fluid, caused by collisions with fast-moving molecules in that fluid. It's a statistical phenomenon, first observed in 1827 by Robert Brown, a botanist, while looking at pollen grains in water under a microscope. Brownian motion is not to be confused with Brownian gas, which are plasmas exhibiting stochastic particle behavior and considered highly speculative 1950's Russian propaganda.

The key challenge is that creating such asymmetric interactions while respecting the second law of thermodynamics is extremely difficult—unless one has a specifically compartmentalized 'need-to-know' and exceptionally high security clearance.

Maybe the inner space of a UFO is comfortably maintained by a vortex of neutrinos, which can be terminated with a counter-rotation—a bit like a self-contained super magnetic field—but if this were the case, and we have such magnetic fields with *keeper bars* (where removing the bar collapses or disintegrates the neutrinos), it would appear A LOT less costly and much more efficient than building uber expensive red herring neutrino detectors around the globe. I suspect that this possibility would also resemble plasma or ball lightning—and neutrino detectors are built for very different reasons than we're being told.

Alas, anecdotal evidence and educated conjectures—again the stuff of fiction—may suffice to establish reasonable hope in matters of space exploration, established paradigms, or accepted science—but are more likely to spawn reasonable doubt and discredit with the help of quite effective propaganda machines. Even way back in 1991, more than thirty years ago, a man named Stan Deyo casually mentioned ball lightning otherwise known as *coherent* stabilized plasmas provide unlimited green energy sources. Again, Tesla knew too.

Science, government, and religion are notorious for vehemently playing a game of reject, dismiss, and deny when it comes to reverse-engineered anomalies, advanced technology, confiscated patents, and supernatural phenomena, labeling them unexplainable flukes or nonexistent specters of conspiracy theory—or even easier, rendering such topics as the realm of the devil.

Useful intellectual advancements in science are shrewdly eradicated from the public domain or ignored, as Bohm and Tesla's examples of ostracization remind us. Additionally, having friends in the industrialized-military-intelligence complex helps to keep conventional status quos and truth crusaders reading heavily redacted straight lines.

From the first day scientists sign up for work on clandestine projects, they must sign papers, under threat of prison and steep fines, to keep classified projects secret. The Espionage Act in the US and the Foreign Secrets Act in the UK are but two examples of how much *heat exposure* and career radiation anyone will face for violations of either act.

When such anomalies aren't overlooked, tactics like ridicule, intimidation, and discrediting campaigns are frequently employed to suppress innovation and to obscure the accidental exposures of clandestine rituals or covert projects—and then filed away and deeply buried under national security concerns, forever lost to public eyes and scrutiny.

Typically, the classified technology is then fed to cronies or watered down and passed off in pieces to private corporations, while original inventors get screwed out of their rightful royalties, credit, and compensation. What can I say? MICE (money, ideology, compromise and ego) motivations are an effective and mainly successful seductress, alongside fear, bankruptcy, and the threat of prison.

Like the existence of dark energy, proving underhanded dark matters is virtually impossible. The militarized-industrial-intelligence complex is a well-oiled, tightly controlled, highly functioning quantum machine in and of itself.

Raging against such a deeply compartmentalized system, which weaponizes intelligence agencies and law enforcement, promises frozen assets, accusations of treason, lost jobs, reputation destruction, and in worst case scenarios, prison, serious injury, or death to oneself or loved ones. It is a practically applied and useful deterrent.

But let's face it—even the best whistleblowers are but a drop in the ocean of secrecy, risking it all (and others) to offer mostly confusing scraps of incomplete information due to something called *compartmentalization*. A whistleblower does little to advance any sort of meaty disclosure regarding anything of high-level value and his or her information, once public, is put forth as more of a public distraction.

To counter whistleblowers, the *sheeple* are occasionally fed a mix of misinformation with truthful tidbits and always promised a gourmet meal of complete technological or extraterrestrial disclosure...later. Later never arrives. Mostly, and in place of disclosure, we are *wowed* by some new technology or propaganda that's existed for years in the covert sector—and this effectively pacifies or severely dampens our lack of *need-to-know* until the next half-secret comes along.

And what I've never understood, given some of the sworn enemies we allegedly face, is why we'd kill off our best minds even if they did blow a whistle? Much like black hats in cybersecurity, we should be giving them promotions, using every drop of knowledge they can provide for our national security—or if our egos and indignation are too big for that—unceremoniously discredit them, fairly compensate them to go away, and let them move on with their lives.

Da Vinci left Italy for the first time when he was in his sixties. He left because his genius was no longer in demand, and he grew weary of bureaucracies, politics, and boring commissioned works. On an invitation from King Francis I, da Vinci was given the title, "Premier Painter, Engineer and Architect to the King" and lived at the Château du Clos Lucé, with his entourage—and at his leisure. The king had no expectations of da Vinci—he greatly admired the sage and encouraged da Vinci to *play*.

During a private audience one afternoon over tea, King Francis inquired about what artistic or thought endeavors da Vinci had brought with him from Italy. "The Premier Painter, Architect, and Engineer to the King" produced the artworks of the Mona Lisa, Virgin and Child with Saint Anne, and Saint John the Baptist. Francis thought the works magnificent.

And many Italians, even today, feel that France stole the Mona Lisa from Italy and have, without success, demanded its return. Da Vinci's alleged lover, Salai, it is said, fairly and squarely sold the painting to King Francis after da Vinci's death—much to the consternation of Italians everywhere.

Da Vinci's departure from Italy and leisurely end days in France, hanging out with, and being fully supported by a King, is likely one of the greatest examples of a human being showing up where they feel wanted and appreciated —and leaving behind an astounding legacy that is still fought over. The same might be said for other great scientific and spiritual talent we choose to cut loose or otherwise unduly dismiss, discredit, or chalk up as crackpots or heretics.

Consequently, I fully acknowledge that national security is a valid and vital reason for maintaining secrecy. It plays an indispensable role in safeguarding a nation's interests. While humanity collectively aspires to global peace and an end to relentless conflict, there will always be those who seek to harness advanced technologies, violence, and terror in their quest for domination. This harsh reality highlights the urgent need to cultivate unconventional, visionary thinkers—individuals capable of countering such threats with creative, peaceful solutions. Yet, the very secrecy designed to protect us can just as easily become a breeding ground for unchecked power and systemic abuse.

Ultimately, national security and counterintelligence require a delicate act of harmonization. And as for the idea of a 'need-to-know'—often, the truth is we're better off *not* knowing—sometimes for the sake of our sanity and safety.

Nikola Tesla's visionary experiments with wireless energy transmission were initially and vehemently dismissed by mainstream scientists (and the military) yet paved the way for technologies like Wi-Fi, laser weaponry, and wireless charging.

Competition in this realm was fierce and men such as Tesla were often stalled and stifled in favor of establishment order. This lack of vision, in my unimportant opinion, sets some governments way behind others in advanced technology as we'll soon see.

Similarly, Dr. Ignaz Semmelweis, who once advocated for something as simple as handwashing to prevent the spread of disease, faced outlandish ridicule before his ideas became foundational to modern medicine.

These examples underscore the importance of adhering to logic and scientific rigor, while also daring to dream and to question, boldly and repeatedly, established norms. Dusting off what we can't explain prompts us to stop pushing aside curious anomalies out of fear or ego and *EXAMINE* them.

I realize that going against scientific comfort and convention isn't easy to do. Humans tend to operate under a predominate negativity bias—a childlike and primal psychological phenomena where we are influenced to lean toward 'the bad' (status quos) and forget the positive (innovation). We consume our existing talent over further developing it—and therein lies our primal shame.

Part of this stems from the fact that bureaucracies typically fund mainstream projects and frown upon alternative sciences and free-thought—and this is part of the reason they grow stale. If everyone keeps their head down trying to get to retirement, there is no more service or advancement *"for the people"*. As Bohm suspected, knowledge becomes static.

Big government is unlikely to change because it's also rife with waste, corruption, and territorial disputes between agency and political leaders. And I would postulate, its loaded with sociopaths (an environmentally learned trauma) —and likely a few psychopaths (a genetic condition).

Alternatively, in areas like the supernatural, where empirical evidence is elusive, the willingness to entertain the imagination might lead to breakthroughs in understanding phenomena that defy explanation, pushing humanity closer to unraveling the core truths of our quantum existence. By balancing reason with open-minded curiosity, we expand the boundaries of what is possible.

Switching gears here, I remain unconvinced about the Big Bang—what if the beginning of the universe was more like a zinc spark comparable to what happens when a sperm meets an egg—and then it divided from another universe, like a cell?

What if our universe is the multi-billionth split of that cell—and other worlds just like ours, soldier on? It's almost too much for any scientist or religious soul to consider, that our bubble of a galaxy, or even multiverses, might be *cells*, a miniscule part of a ginormous living *something else*.

Consequently, researchers at a Northwestern University-led interdisciplinary research team that included experts from the U.S. Department of Energy's Advanced Photon Source at Argonne National Laboratory, proved back in 2014 that zinc joins itself to small, light-emitting molecule *probes*.

The zinc spark is a direct reflection of the "egg's" quality and ability to grow into a viable embryo. If we can do this in a petri dish, I'm sure the universe could divide itself using light sparks on a much grander yet less explosive scale that should have obliterated neutrinos but didn't.

In 2019, *Astrophysical Journal* published a paper where researchers from MIT reported a strong abundance of zinc in, HE 1327-2326, an ancient star and *second-generation* star, among many. They believe the star could only have acquired such a large amount of zinc after an asymmetric explosion of one of the very first stars had enriched its birth gas cloud. And no matter how they measure it, an abundance of zinc exists.

Interestingly, zinc is used to prevent steel and iron from rusting and Zinc-Oxide coating is routinely applied to advanced space technologies to prevent corrosion in space. It is also the fourth most used metal on the planet. Zinc, it seems, is crucial for expanded life.

Congruently, Hermes Trismegistus said, *"That which is Below corresponds to that which is Above, and that which is Above, corresponds to that which is Below, to accomplish the miracles of the One Thing."*

His phrase appears in the Emerald Tablet of Hermes Trismegistus, an ancient document that originated between 200 CE and 800 CE. If he is right, humans are indeed, one thing, composed of star stuff.

Climb in, we're going for a ride into the spooky-shivers side of matter and energy…

CHAPTER SEVEN

A Quantumly Conscious and Subconscious Universe

Ever wonder what happens when quantum physics and consciousness meet before you've had your morning coffee? Picture this: you're sitting in your favorite chair, uncertain if you exist, when suddenly this book informs you that you're a wave of infinite possibilities who's been living in a super positioned state.

As if existential crises weren't bad enough, you now grapple with the fact that the very atoms making up *you* and your coffee might be entangled with another cup of coffee simultaneously poured in Orion's belt.

And that's not even the weird part—the weird part is that by simply looking at your coffee, or anything else around you, you're forcing countless quantum possibilities to collapse into one caffeine-infused routine reality.

Welcome to the bizarre intersection of quantum mechanics and consciousness, where Schrödinger's cat isn't just alive and/or dead—it contemplates its existence, considers switching to decaf, and simultaneously hopes for catnip.

The Nature of Reality

From a quantum physics perspective, reality at its most fundamental level behaves in ways that challenge our intuitive and analytical understanding. As already discussed, the quantum world exhibits properties like *superposition*, where particles exist in multiple states simultaneously until observed. And in *entanglement*, where particles can be correlated across vast distances in ways that seem to defy classical physics.

This creates an interesting tension between the macroscopic reality we experience, which appears solid, continuous, and governed by classical physics—and the quantum realm, which is probabilistic and mostly counterintuitive.

As a refresher, some interpretations of quantum mechanics, like the Copenhagen Interpretation, suggest that reality only takes definite form when observed, while others like the Many-Worlds Interpretation propose that all possible quantum states exist in parallel universes.

The relationship between consciousness and quantum physics is particularly intriguing. While some theorists have proposed that consciousness might play a role in "collapsing" quantum wave functions into definite states, others argue that quantum effects are effectively cancelled out at the macroscopic scale through *decoherence*. Beyond physics and science, there's the philosophical question of whether our perceived reality is fundamentally "real" or if it's more accurately described as a model constructed by our brains to help us navigate our environment—or we're not real—and a holographic reflection of something real—or we're perhaps living in a simulation.

Our sensory systems allegedly only detect a narrow band of available electromagnetic radiation, sound frequencies, and other physical phenomena, suggesting that our experience of reality is inherently limited and extraordinarily filtered. We not only have the wool pulled over our eyes—most of us spend a lifetime wearing rose-colored glasses too! At any given time, we *think* we're seeing the entire landscape in front of us—but it's only a fraction, while our brain fills in gaps on life's canvas—a *perception* based on *illusion*. A brane game within a brain.

This inability to *see* beyond a 3-4D reality connects to questions about the hard problem of consciousness—how and why we have subjective experiences at all, and whether consciousness itself might be a prolific permeating fundamental part of reality rather than an emergent property of complex systems.

What is Consciousness?

Consciousness, we're told, is a special sort of subjective awareness and first-person experience through the senses and mind. Philosophers sometimes call it "qualia" or "what it feels like" to experience something. We can laundry list a host of words to describe consciousness, but thus far, humans are unable to succinctly define it, test for it, measure it, or precisely locate it in living (or dead) biological brains. While we can observe neural correlates of consciousness within the brain, explaining how physical processes give rise to subjective experience remains a major scientific challenge known as the "hard problem of consciousness."

Regarding the role of consciousness (or not) in quantum physics, there are several perspectives to consider.

The measurement problem in quantum mechanics raises interesting questions about consciousness.

The Copenhagen interpretation suggests that quantum systems exist in superposition until measured/observed, which led some physicists like Eugene Wigner to propose that consciousness itself might play a role in "collapsing" the wave function. Most modern physicists shrink from and are highly skeptical about lifting consciousness into such a central role.

An alternative view is that consciousness emerges from complex information processing in physical systems (like brains) that follow quantum mechanical laws but doesn't play any special role in quantum phenomena. This is more aligned with current, materialistic scientific thinking.

Some philosophers and scientists have proposed more radical views, like integrated information theory (suggesting consciousness is a fundamental property of reality) or quantum theories of consciousness (proposing quantum effects in microtubules contribute to consciousness). These remain highly speculative in most accepted fields of science.

What's particularly intriguing is how consciousness seems to create our lived experience of reality, including our perception of time, space, and causality. Yet we still don't understand if consciousness is merely an emergent phenomenon from physical processes, or something more fundamental to *reality* itself—maybe something emerging or existing in the quantum *splooge*, which moves through the entirety of space the way neutrinos so easily pass through matter.

Physicalists studying biology or brain tissue typically refuse to jump from petri-dishes into deep consciousness fryers. They firmly grip the beaker of a more comfortable reality, which claims that the brain only operates at the classical level of physics (neurons firing, proteins reacting, receptors receiving, electro-chemical impulses etc.).

For materialists, the human brain is a spongy meat-like substance that scientists and doctors divide into elementary categories, where each section or structure is responsible for a function or behavior, one that communicates and integrates networks to sustain and propel a functioning sentient being.

According to materialist philosophers and neuroscientists, consciousness emerges from brain activity as a product of neural function. This view holds firm that consciousness cannot exist independently of a functioning brain. However, this 'the-brain-is-meat' position struggles to explain several well-documented cases where individuals with severe hydrocephalus—resulting in dramatically reduced brain tissue—have displayed normal cognitive function and consciousness.

For instance, research published in Science in 1980 detailed a case of a 44-year-old man with an IQ of 126 who lived a normal life. He was found to have less than 10% of the expected brain volume. Scans revealed that most of his skull was filled with fluid, leaving only a thin layer of brain tissue. He had suffered from hydrocephalus (a condition where cerebrospinal fluid accumulates in the brain) since childhood, which had caused his brain to compress. Despite his condition, he held a job as a civil servant and was married with children.

Such cases, while rather rare, pose significant challenges to strictly mechanistic theories of consciousness and shows us that our understanding of brain structure as it relates to conscious experience remains woefully incomplete. Maybe, as we've just read, a compressed brain works just as well as an inflated one. The wiring and energy are still there—it's just compacted.

Another case in point is a 20-year-old college student in the UK, found to have virtually no brain tissue when doctors scanned his head due to unrelated concerns. His condition, also linked to hydrocephalus, resulted in only about 10% of the expected brain tissue. Despite his physical condition, the student excelled academically, achieving honors in mathematics and being described as a "completely normal" individual by those who knew him.

Some theorists like Sir Roger Penrose, whom we'll get to in a moment, propose that consciousness might involve quantum processes in microtubules within neurons of the brain. There's some evidence supporting his idea, but it gets unfairly relegated to the orbit of pseudo-science. I suspect this is likely due to both the rigors imposed by the scientific method, highly nervous and biologically adverse tenured lecturers, and the fact that expanding these concepts beyond the strict scientific paradigm might collapse the need for science and put a lot of people out of business and careers.

To be circumspect, most neuroscientists remain skeptical of Penrose's theory since the brain appears too "warm, wet, and noisy" to maintain *quantum coherence*. But what if microtubules, like UFO's or other aerial phenomena, are closed systems, ones capable of completely shutting out the warm, wet, and noisy?

Harold Saxton Burr, a biologist, visionary, and high-thinker integrated physics and philosophy into his work. He closely dealt with closed system patterns of self-organizing consciousness through bioelectric ignition. And he once wrote a paper with F.S.C. Northrop entitled, *The Electro-Dynamic Theory of Life*.

Moreover, Burr was fascinated by the concept of *bioelectricity*, naturally produced voltages within living tissues, and how it correlated with important nonlocal morphogenetic and patterning events in molecular and atomic constituents of protoplasm.

Burr viewed electric potentials as lattice that guides cell behavior. This reminds me of a hybrid circuit board—a concept elegantly connecting biology to technology—suggesting cells function as microscopic electrical systems with their own programming to maintain life's harmony.

Subsequently, he correctly discovered that cancer tumors can be pre-detected through bioelectric pattern disturbances—which leads me to speculate that if they can be detected, they can also be eradicated.

Burr knew this more than seventy years prior to the discovery that nonlocal imposition of bioelectric states by optogenetic techniques can *normalize* cancer! So, what the hell does that last sentence mean? It means that he knew we could "reprogram" cancer cells by using light. We can send signals into cells that reset their electrical activity—and we can do so using a belief system.

This is like fixing a glitch in a computer. By doing this, cells 'gone wild' have the potential to return to their normal, healthy behavior, without being directly attacked by radiation.

Then, some research came out in 2023, where chemists Aarat Kalra and Gregory Scholes out of Princeton, led a study into how light energy absorbed through microtubules produced surprising results (tryptophan autofluorescence lifetimes to probe energy hopping): the light energy diffused almost five times beyond boundary expectations, according to their classical calculations. This suggests quantum resonance phenomenon in the microtubules. Ultimately, quantum effects in the brain, do, on some level, persist.

Backing up by only a few years, in 2018, Na Li, at the Huazhong University of Science and Technology in Wuhan, China, conducted a study on eighty anaesthetized mice using xenon gas isotopes. Simply put, the isotopes contained an odd number of neutrons in their nucleus, giving them a quantum property called "spin", which allegedly weakened anesthetic effects by almost twenty-percent. Notably, spin makes nuclei act like a tiny bar magnet and can only be explained through equations of quantum mechanics. Li later said that this anesthetic spin suggests that consciousness relies on quantum phenomena.

Thankfully, given these promising results, other skeptical scientists are now jumping on the proverbial quantum brain bandwagon.

As we've discovered, we might make all the plans we please, but seldom, does anything go **exactly** as planned, researched, or imagined—with life, quantum mechanics, science, religion, war—or with our brains. At some point in time, somewhere, change happens, no matter how much we might try to contain it.

Additionally, most humans move through life on autopilot, consumed by matter that doesn't matter—or insisting that matter is all that matters. How many times have we needlessly ruminated or worried about *something* only to discover, as distance is created between us and whatever it was, we once craved, suddenly holds much less value or urgency?

Consider your Christmas wish list when you were a kid. Are you still convinced you'll just die if you don't get that tricycle you so desperately wanted as a four-year-old? What about when you were a teenager, when you endlessly obsessed over a mad crush, thinking the world would end without this other person, your soul-mate, reciprocating your earth-shattering, life-long true love—what if that very same person showed up today to insert themselves into your life? And now that you're an adult, would you trade your mode of transportation for a tricycle?

Human beings control nothing yet *think,* with our brains, minds, senses, and hormones, as if we control everything. And this is what happened to this book. I took a wild detour into the mind, matter, and how quantum particles and that sticky-wicket of a substrate field might impact us a lot more than we ever consider.

It's true that humans generate unique forms of frequency, energy, or vibration—positive or negative—and send it out into the world, to return to us as treasure or trash—but the reality we end up creating is usually far from an exact replica of our fantasies. This is largely because complex systems, such as other humans, have their own energetic inputs on events and environments, counter-influencing our realities—but not as much as our subconscious mind does a personal number on each of us. More on that in a minute.

Reality, like time, is a mind construct to make sense of energy fields posing as matter and its changes, which, behind the deepest veil of existence, is an illusion made up of waveform particles above (or around) a spongy sort of vibrational substrate—and we put way too much emphasis on and worry into, the illusion, while rarely, if ever, considering the vibratory fields or foamy *splooge* underpinning subatomic sentience.

However, we do use our *conscious* mind to label, categorize, judge, build, and pragmatically order that which ultimately lies beyond human comprehension. This is done so we may productively and reasonably live in and make sense of our 3-4D world. But we, as humans, are oh so susceptible to quantum manipulation and outright deception.

For example, it is one thing to claim that a concrete block is made up of vibrating atoms that never touch one another, and is therefore, only giving off the *illusion* of being a concrete block when we observe it—and quite another thing, in our 3-4D world to *feel* the exact same concrete block fall on our foot. And—if we don't pay our bills because we consider them *illusions,* the electric company or our mortgage broker will, eventually, shatter this illusion when we find ourselves living outside in the dark. Perception and fear are great drivers of linear thinking, arrows of time, hard work, and conformity.

Admittedly, without a logical, conscious mind, maneuvering effectively and efficiently through 3-4D sentience would be difficult if not nearly impossible for most humans. It is rare to find any individual capable of physically moving through solid objects unimpeded without the use of specialized technology or an exceedingly well-trained perception system. Regular, everyday human lives seem to require a certain level of epistemic or instrumental rationality (and morality) to minimize environmental, personal, and physical pain or discomfort.

Correspondingly, while the conscious mind helps humans present themselves to the world as largely responsible adults, it is the often child-like *subconscious* mind that majorly hobbles human existence (and transcendence), while at the same time making life on Earth appear sensical. The subconscious mind is a foundational undercurrent of our biology, housed, I'm convinced, within our DNA. It is the location of our programming and reticently rules over most humans without their awareness.

During the 1990s, scientists collaborating with the U.S. Army conducted an intriguing experiment involving DNA samples from volunteers. These DNA samples were placed in one room, while the volunteers remained in another. The participants were shown a series of emotionally charged videos designed to elicit strong emotional responses. Remarkably, as the volunteers experienced these emotions, the DNA samples in a separate room exhibited significant electrical responses, as though still connected to the individuals' bodies—which makes me think that on some resonance level, this is how many mothers might sense something wrong with their children, who may be miles apart from them.

Dr. Cleve Baxter, a scientist with ties to the CIA, took the experiment a step further by separating the DNA from the participants by a staggering two-hundred-fifty miles. Astonishingly, the DNA still reacted instantaneously to the participants' emotional states, suggesting a hair-raising possibility: DNA might electrically, through quantum processes, instantaneously superposition or entangle us in ways science has yet to fully understand.

Furthermore, whenever we touch someone, we exchange or leave traces of DNA behind. Depending on the environment, this DNA can survive for days, weeks, or even months. Could this suggest the existence of an invisible energy field that binds us to others as long as our DNA remains viable? This concept opens fascinating questions about Bohm's hidden variables of interconnectedness and the potential influence of DNA beyond physical proximity.

While not directly related to DNA, another example involves human blood: Dr. Rebecca Marina studied her own red blood cells under a darkfield microscope (a special type of microscope used for the study of small particles not visible under ordinary lighting). The photos showed her blood before and after she used *Emotional Frequency Therapy* to treat different emotional states. She says she was inspired to do this by the work of the Japanese researcher, Masaru Emoto, who photographed pictures of frozen water crystals imbued with different emotions.

When Marina felt sad, her blood cells became tear-shaped! If she experienced tension or stress, her blood cells clumped together. But when she felt love, her cells turned back to normal and even sparkled with tiny lights. Shockingly, when scientists prayed or otherwise set an intention over her blood cells (cells separated from her body for hours), the blood cells started glowing and pulsing at the same rate as her heart, while the cells receiving no intention didn't change! This is phenomenal.

If an individual's blood cells can glow and pulse with light under *external* or self-induced intentions of love, might they also be ripe for abuse if the intention is reversed and negative? From a national security standpoint, could blood cell or DNA manipulation using external ill intent directly impact a person's mental and physical health? As the previous two scientific examples show, the answer is an astounding and frightening *yes*.

Alternatively, but still connected to DNA and blood, the subconscious mind is a great thing because it enables us to ingrain repeated practices into automatic memory—it is the reason we're able to drive a car, ride a bike, dance, read, play an instrument, or make up our bed with military-grade precision, without thinking too much about it. It provides us with what some have termed *automatic flow* through practice and repetition.

But the subconscious mind also deeply buries trauma and often runs on a child-like operating system. It does this to help us function and get on with life after a horrible event, much the way a tumor corrals cancer to protect the rest of the body—but somehow the trauma embeds itself into our very being (and maybe our blood and DNA), even if we're no longer able to identify or remember its details—and it rumbles around inside us—that *thing* of unease we sense but are often helpless to willfully uncover or counter—unless we receive medical intervention, spiritual training, or alternate therapy. The subconscious mind might also make us unwittingly subject to ill intent—unless we counter intent through something I term bio-spiritual shielding.

Our subconscious mind and its unaddressed traumas keep us stubbornly tethered to harmful habit patterns or behaviors such as impulsiveness, co-dependent relationships, addictions, or even block the ability to stick to a budget or diet. Bad habit patterns and/or trauma literally embeds into our cells, which also store memory and emotion, and exchange information in the cell membrane.

If we fail to uncover that which subconsciously drives us, we become slaves to it. It helps to understand that physically, our brains develop networks—almost like electrically charged cable wires—that can be connected or disconnected into bundles and rewired for greater function or dysfunction. We do this to ourselves without realizing it, or others sometimes do it to us, especially if we focus attention, energy, and perception toward them.

Subconscious programming sort of works like a curse might; the level of a curse's success depends on the focus and attention we give it. If we do not believe in curses, they have little, if any, impact. If we believe in curses, even subconsciously, they will likely be effective. We can also use the subconscious to deflect such attacks through disconnection, meditation, hypnotherapy, bio-shielding or reflection-refocus.

Fortunately, with practice and awareness, sentient beings can transmute negative energies of the subconscious mind and positively channel it through and out of our cells. Humans can rewire brain circuitry too, known as neuronal networks, to build up or shut down certain helpful or disruptive patterns the way we might build muscle, balance, or stamina, through exercise. We *can* effect quality changes in ourselves or others by learning about our subconscious and how it works, mostly undercover, on our biology and behavior.

Consider headaches, anxiety, panic attacks, IBS, and ulcers. These events, we're told by doctors and researchers, are triggered by stressors. The stressors cause largely pent-up mental anguish, which is often manifested by the undigested sufferings of the subconscious mind and then stored in our cells. Our negative thoughts and subconscious emotions then manifest physically, contributing to measurable medical conditions like tumors or heart attacks. This represents a powerful form of energy conversion—one we often create unknowingly as we store unprocessed emotional pain in our bodies.

To better understand and manage this mind-body connection, we might practice mindful body scanning to notice physical tension and emotional patterns or work with qualified therapists, doctors, and with other healing modalities to safely process stored trauma.

We could develop regular meditation or contemplative practices, journal our emotional experiences and physical sensations, or study ancient traditions and evidence-based mind-body healing approaches. And some of the best healing techniques for an overloaded subconscious mind is exercise, prayer, chanting, or breathwork.

Besides that, energy, insofar as I know it, is power derived from the utilization of photons to create light, heat, or movement—to construct, to build, to manifest or destroy—and it comes in a variety of forms. The color of a photon also determines the sort of energy it carries. All space is filled with energy. If we can adjust it for craft propulsion, it can also be adjusted for personal transcendence.

It does appear, based on the elementary quantum lessons we've learned in previous chapters, that we can set our minds upon all sorts of matter to effect change. We can *observe* our choices, conditions, attitudes, thoughts, health, or circumstances—and through a conscious focus—collapse vast possibilities of the subconscious into a single decision, positive or negative, based on the energetic current we send through our neuronal networks, DNA, and cells, and out into the universe. We can also create something physical from an idea, and we can harmonize our blood and bodies to the frequencies upon which we set our intentions.

On top of that, beneath quantum mechanics lies *something*, which cannot yet, we're drilled to believe, be measured or observed—but it can be harnessed. It is intelligent and responsible for animating everything we know into existence—and for making our lives a living Hell or a paradise on Earth—based again on our individual *perception.*

Once humans learn to recognize how they are influenced, can influence, and are predominately caught in a matrix of unconscious manipulation and deception, many awaken and consciously flip the switch to bring light into their lives.

Perception is the one thing human beings have the power to change—and in doing so, we dismantle or turn off neuronal networks that no longer serve us and turn on the ones that do. This is a remarkable ability! And I'm convinced neutrino-type vortexes within quantum fields are key to helping us lead incredible lives through harnessing vibratory fields. This quantum substrate of force is beyond the grasp of most earth-bound, biologically contained humanity. But it doesn't have to be.

Unfortunately, most humans have been indoctrinated through institutions, epigenetic memory, and conventional subconscious programming into an elementary form of semi-awareness and auto-pilot—one where we've lost touch with nature and rely too heavily on authority figures, money, and technology to fix us, instead of taking the lead role in our own personal care and accountability. We fail to consider quantum processes that greatly influence *being* by becoming completely dependent on and tethered to technological amenities, other humans and/or instant gratification.

Likewise, it's true that humans, like other animals and plants on Earth, must have sunlight, water, food, and air to *live.* Living on this planet requires the exercise and interconnectivity of energy and consciousness to both survive and thrive. And no matter how high we elevate our mind, the body must always have the basics of sentience to survive, or a human must transform into something else and transition off planet.

Interbeing with the world on a physically conscious and subconscious level and getting on well with our individual and collective realities, requires harmonic *interdependence*—aka coherence—aka wholeness. It is realizing that humans are the best and worst stewards of the resources of this planet—and its only hope. If we destroy this place, where do we go? I'm unaware of any other planet within our solar system better equipped to host human life.

Energy makes up both mind and consciousness, which are considered distinct but interrelated. The mind is a cognitive system that processes thoughts, memories, emotions, and sensory information. It includes conscious and unconscious mental activities such as thinking, reasoning, flow, and decision-making.

One's mind is also associated with brain function, yet a debate rages on whether the mind is purely a byproduct of physical processes—or a combination of biology and something quantum that lies outside of the human body but works in tandem with our biology.

Interestingly, piezoelectricity is the body's generation of electric charge in response to mechanical stress. In the human body, bones, tendons, fascia, and collagen exhibit piezoelectric properties. Piezoelectric materials can also be used to harvest energy from movement to power electronic systems.

Like something out of a sci-fi novel, recent advancements in brain-computer interfaces (BCIs) have enabled the control of drones using neural signals. A notable example is the "Brain-Drone Race" held at the University of Alabama in April 2023. In this event, participants wore electroencephalography (EEG) headsets that detected electrical activity in their brains, allowing them to control drones through thought processes. But that's not all!

In 2021, a UK company called Ultra Electronics unveiled a "brain piloting interface" that enables users to fly drones using only their minds—without any wires! The system works by having the pilot look at specific icons on a computer screen, which elicit unique responses in the visual cortex. These responses are detected by a sensor and translated into commands that control the drone's movements. This is exciting for technology advancements—and potentially frightening for future wars.

As previously stated, energy is the fundamental capacity for action or change. In living systems, it manifests as biological energy (ATP), electrical activity (neural impulses), and electromagnetic fields. Some ancient wisdom traditions and practices suggest subtler forms of "life force" energy, or a quantum unseen flow that some humans regularly tap into to realize greater harmony, higher awareness, and health.

Life force energy, known as "qi" (Chinese), "prana" (Indian), or "ki" (Japanese), is considered vital for animating living beings. Some ancient and non-Western medicine systems view this energy as flowing through pathways in the body (meridians/Nadis) and believe personal health depends on its balanced flow.

While life force energy itself isn't scientifically *proven*, because it is impossible to measure such unknowns or subjective experience at a quantum level, several alternative practices show measurable benefits in research:

- **Qigong:** Studies indicate benefits for balance, flexibility, blood pressure, reduced inflammation markers and chronic pain management

- **Acupuncture:** National Institute of Health (NIH) studies demonstrate benefits for migraines, osteoarthritis and chronic pain management. Research suggests impacts on inflammatory markers and nervous system function
- **Reiki:** There is some evidence for pain and anxiety reduction and NIH studies show possible benefits for relaxation
- **Meditation & Breathwork:** Harvard studies show meditation reduces stress markers. Other research demonstrates improved immune function and brain imaging scans shows structural changes with regular practice
- **Music, Chants, Prayer, and Dance**: The (not-so) secret code of the universe

Modern interpretations sometimes link life force energy to measurable bioelectric fields or cellular energy production, though this differs from traditional understanding of life force as a subtle, *non-physical* energy. And many scientists and medical doctors have stated, quite assuredly, that alternative modalities or medicines are risky, and any benefit allegedly derived from them is likely the result of the placebo effect. Hold on. What is, exactly, the *placebo effect*, and if it works, what causes it to work?

Briefly, the placebo effect is a beneficial health response triggered by a person's *belief* in a treatment, even when that treatment has no active therapeutic properties or scientific basis in modern medicine or pharmaceuticals. It is a scientifically well-documented *phenomenon*, which produces highly measurable physiological changes.

The placebo effect works through unquantifiable *expectations* that trigger the release of endorphins and dopamine, natural pain-relieving chemicals in the brain, which provide relief to the person who *expects* relief. If the expectation of relief is met, this sets up something called *learned response*, where a successful *treatment* (the placebo) will reactivate or be triggered due to *psychological* conditioning.

Belief in a non-treatment (or God, or even superstitions and curses) literally alters how a person interprets their symptoms (and their experiences) to realize biological healing and elevated *first-eye* abilities. This is incredible! This is proof that some form of unseen energy is harnessed by the mind to give rise to biological and mental healing. Used in reverse, or flipped inside out, humans, as we learned earlier, also do this to make their lives a living hell full of drama, sickness, and suffering. We are, according to the placebo effect, our own voodoo queens, Gods, or healers.

The placebo effect varies by condition and belief level. According to research, it appears to be most effective for pain, depression, fatigue, anxiety and nausea. Yet some Chinese *thought* surgeries have also cured patients of cancerous tumors, which shrink and disappear on screens as the patient is having hands laid upon them without use of surgical tools!

Christians do virtually the same things when they collectively lay hands upon people and fervently pray to heal them. The level of healing a recipient receives seems to coincide with levels of their readiness to believe and their *faith*.

Consequently, it is interesting to note that a few medical professionals are now looking at cancer tumors, not as body invaders bringing illness, but as harvesters of sickness that corral abnormal cells into lumps and bumps designed to protect the rest of the body—lumps and bumps that could be shrunken or exorcised from the body using a hybrid of prayer, placebo-based, and modern medical practices.

Physical manifestations of the placebo effect based on research include reduced inflammation, altered heart rate, and improved immune response. Notably, placebos work even when patients *know* they're receiving them, suggesting the effect isn't solely based on deception, but on much broader psychological and neurological mechanisms. Could the placebo effect be a highly-tuned vibrational intention where the patient and participants experience shaman-like healing?

Conversely, the nocebo effect happens when a person experiences a negative outcome from a harmless treatment (or curse, or spell, or words) simply because they *believe* it's going to harm them.

The problem is that scientists are unable, at this point, to exactly pinpoint these mechanisms, what sets them into action, or how they work. In the instance of a placebo or nocebo effect, each are proven to exist, but like quantum physics, it flies in the face of empiricism, logic and common sense—three essentials that seem to limit the scientific mind and hinder established paradigms. The nocebo and placebo, like quantum physics, is a paradox.

Furthermore, *belief* is a mental state of acceptance or conviction in the truth of an idea, which triggers cascade effects throughout the body via the nervous and endocrine systems. This mind-body connection operates through several pathways, including human blood and DNA, as Dr. Marina and Dr. Baxter previously showed us with their experiments.

Beliefs activate specific brain regions or neural pathways, releasing neurotransmitters and hormones that affect physiological functions. For example, belief in safety, the power of prayer, God, or healing, reduces cortisol production, while belief in threats or insecurity increases it. Positive beliefs also, studies show, enhance immune function through reduced stress hormones and increased production of antibodies. Conversely, negative beliefs suppress immune responses.

Sustained beliefs create new neural pathways, physically altering brain structure and function over time—and scientists refer to this as neuroplasticity, the brain's ability to restructure itself biologically, accounting, in many cases, for some astounding conscious and subconscious behavioral and psychological changes in people.

Additionally, belief modulates pain perception by affecting how the brain processes pain signals. This explains, in part, how placebos combined with belief might provide genuine pain relief. The biological effects of belief are measurable and reproducible in laboratory settings, demonstrating that subjective mental states, our *perceptions,* have objective and measurable physical impacts. Humans can and do control their worlds and dimensional realities.

CHAPTER EIGHT

The Biology of Quantum Consciousness

Current quantum research suggests several processes for the effects of belief. For example, *quantum coherence* in biological structures, particularly microtubules in brain neurons, might, a few suspect, be influenced by consciousness states. One theory is, as of this writing, highly controversial in scientific circles because how one *feels* is an invisible state that can be *translated* but not reduced through numbers.

Microtubules and Theories of Consciousness

The Orchestrated Objective Reduction (Orch OR) theory, proposed by physicist Sir Roger Penrose and anesthesiologist Dr. Stuart Hameroff, suggests that consciousness emerges from quantum processes in brain cell microtubules.

Microtubules are protein structures in neurons that form the cell's cytoskeleton. Their theory claims microtubules sustain quantum *superposition* states. These quantum states allegedly "collapse" in an orchestrated way to produce conscious experience.

The theory also proposes that consciousness operates at the quantum level rather than just through classical neural activity. It is thought that anesthetic molecules (i.e. endorphins, melatonin, serotonin, enkephalins etc.) work by disrupting these quantum processes in microtubules.

Microtubules, which provide structural support and transport within neurons, likely facilitate quantum effects in the brain. Penrose and Hameroff propose that quantum processes in microtubules could influence neuroplasticity through:

- Quantum coherence affecting protein conformational changes
- Quantum entanglement potentially coordinating synaptic modifications
- Quantum computing-like processes in microtubule networks

However, this quantum connection between the human brain and consciousness remains highly theoretical and debated in neuroscience. The main challenge is maintaining quantum effects at biological temperatures, though some evidence suggests microtubules might provide sufficient isolation for quantum processes, as earlier discussed—and some recent research involving mice under anesthesia seems to back up Penrose and Hameroff's research.

Neuroplasticity

Neuroplasticity is the brain's ability to reorganize itself by forming new neural connections throughout life. It's essential for personal growth, well-being, learning, memory, and recovery from brain injuries. The process occurs through strengthening or weakening connections between neuronal networks—and can be designed to help or harm a human.

Rudimentarily stated, neurons, which have three main parts (cell body, branch-like dendrites, and long fibers called axions) are specialized cells in the nervous system. They release neurotransmitters (brain chemicals like dopamine or serotonin) from synaptic terminals to communicate with other neurons.

Neurotransmitters cross the synaptic cleft and bind to receptors on receiving neurons. The human brain contains roughly 86 billion neurons, forming complex networks that enable thinking, memory, sensory processing, and movement. Think of it as an electrical circuit board of the human body—capable of upgrades and increased storage capacity and energy—but sometimes, it can spark and sputter in humans too, sort of like the way the way an electrical wire might develop a short and cause a light to flicker or go out—especially if we repeatedly engage in risky or addictive behaviors.

Some of the best ways to foster neuroplasticity, according to scientists and wellness professionals, is to learn new skills, exercise, sleep well, practice mindfulness, make social connections, limit alcohol and drug consumption, and step out of routine to challenge the brain with novel experiences. Consistent meditation practice has also been shown to induce neuroplasticity, leading to structural and functional changes in the brain, potentially improving cognitive function and emotional regulation.

Neural Networks

Neural networks are interconnected systems of neurons that process information through weighted connections, similar to how computers process data. Artificial bio-tech-type neural networks mimic biological brain function but appear—we're told—to lack consciousness.

Moreover, at least in complex biological systems like humans, quantum tunneling may influence neurotransmitter release, entanglement could enable faster information processing across neural networks, coherence might facilitate synchronized neural firing, and quantum effects may impact consciousness and cognitive processing. Neural networks enable learning through pattern recognition, process sensory information, control motor functions, store memories, and allegedly generate consciousness.

While the quantum-biological interface remains theoretical at this moment, it could explain phenomena like instantaneous processing of complex information and aspects of consciousness that classical physics struggles to explain. Recent research, which we'll further explore, suggests quantum effects might be much more stable in biological systems than previously thought, particularly in microtubules and protein structures—or though morphogenic fields.

Prefrontal Cortex Function

The prefrontal cortex (PFC) manages executive functions including decision-making, planning, emotional regulation, and social behavior. It's considered crucial for higher-order thinking and individualized personality.

Cases of individuals living without a prefrontal cortex (PFC) highlight the brain's extraordinary neuroplasticity. Other regions appear to reorganize and compensate, adopting functions typically managed by the PFC through alternative neural pathways and cortical adaptation.

From a quantum physics perspective, it's intriguing to consider whether PFC neurons might leverage quantum tunneling to facilitate neurotransmitter release. Furthermore, quantum coherence could potentially support synchronized neural firing across PFC networks. It's even possible that microtubules within PFC neurons maintain quantum states—perhaps through amplified photonic emissions associated with neural firing.

Whether true or not, it seems that energy transfer involves some sort of quantum processes in mitochondrial networks even if the quantum mechanics of PFC function requires more research to confirm this ability in such a warm, wet, and noisy environment.

The Corticothalamic System

This bidirectional network of brain pathways connects the cerebral cortex to the thalamus and back to provide constant feedback and sensory refinement. Its primary function is to regulate sensory information processing, modulate cortical arousal and attention, and influence sleep-wake cycles. It is crucial for integrating information and conscious experience. Think of it as a highly integrated processing unit.

The corticothalamic pathways consist of neurons that originate in the cerebral cortex (specifically layer 6) and project to the thalamus, which acts as a central relay station. It processes body signals before relaying them back to the cortex.

During sleep, the corticothalamic system is involved in generating rhythmic activity, particularly slow oscillations, which are important for sleep consolidation and memory.

Enhancing the corticothalamic system using specialized technology, increases intelligence, consciousness, and awareness in human beings, potentially greatly increasing overall brain plasticity, according to unsubstantiated research—paving the way for a future of super soldiers, spies, and scientists.

The Substantia Nigra

The substantia nigra is in the midbrain, appearing as paired dark bands of neurons and has two main parts: pars compacta (SNc) and pars reticulata (SNr).

Its role in consciousness centers around dopamine production, critical for movement control and reward-based learning, regulating motor circuits and eye movements through GABA signaling, and contributing to awareness and arousal through connections with the brain's thalamus and cortex

Interestingly, the substantia nigra's dense network of dopaminergic neurons supports quantum coherence through ordered water structures and electron transport chains in *mitochondria* through something like a *quantum dot.*

A quantum dot is the structure on a semiconductor that confines electrons in three dimensions so that discrete energy levels are obtained in electronic devices. It behaves as an artificial atom, whose properties can be controlled. But—quantum dots may also spontaneously form by depositing semi-conductor material (nanoparticles) on a substrate with different lattice spacing (known as heterostructure).

In simpler terms, dopamine neurons of the *substantia nigra* have a distribution and density like photovoltaic devices. Experiments have shown that electron transport is present in the substantia nigra—enough to cause or contribute to the generation of action potentials, which are shown in experiments to alter our perception of time and to control emotion. Maybe the substantia nigra works a bit like miniature neutrino detector cubes about to be mass produced in Germany (stay tuned).

DNA's Role in Consciousness

DNA affects consciousness by regulating neural development, neurotransmitter production, and brain structure formation. It provides the blueprint for building and maintaining the physical substrate that enables consciousness—and it is today, Earth's source code of everything. We are told that DNA acts as the foundation for the biological structures that enable consciousness to emerge, rather than directly contributing to consciousness itself.

Some theories, like the Penrose-Hameroff Orchestrated Objective Reduction (Orch OR) hypothesis previously mentioned, propose that quantum processes in microtubules within neurons give rise to consciousness. But DNA and even blood, as we've seen, might also play a role by maintaining these quantum-coherent structures. This line of research has led me to consider whether reincarnation might stem not from metaphysical phenomena, but rather from epigenetic memory encoded in our blood or DNA—perhaps even from some form of quantum entanglement with ancestral energy fields. These experiences, vivid and emotionally resonant, may *feel* like genuine past lives due to a yet-to-be-understood form of quantum interfacing.

The DNA Consciousness Theory (DCT)

DNA contains a genetic code made up of codons, which are three nucleotides (e.g. ATG and CTG) that, in collaboration with RNA subspecies, ultimately produces proteins. These proteins make up cellular parts and products. Some researchers claim that this is how DNA gives rise to simple forms of life and differing degrees of consciousness, which begins on the cellular level—but possibly sooner in quantum processes.

DCT proposes that DNA molecules possess their own form of consciousness, distinct from human consciousness, which allows DNA to interact with the environment and contribute to the emergence of higher levels of consciousness.

First proposed by researcher John Grandy in 2006, DCT suggests that DNA has a basic level of awareness due to its complex molecular structure and dynamic interactions within a cell, something Dr. Baxter's DNA studies with the CIA seemed to show. For instance, if DNA *electrifies* instantaneously when its original host has an emotional experience, even at a great distance (nonlocality), you can bet there's been both classified research, development, and applications on this topic.

The Sci-Fi Perspective

Writing sci-fi requires imagination and calling upon a variety of inner guidance and outer research, especially as it pertains to blood types, biology, other worlds, advanced technology, consciousness and DNA. Mentally, it's a bit like taking on multiple roles of scientist, philosopher, inventor, dreamer, spy, engineer, and therapist for your characters and worlds. It begins with curiosity and then asking a lot of "what if" questions: What if time worked differently? What if AI had consciousness? What if some covert labs are now far beyond free energy?

Curiosity drives the core of sci-fi to build a believable story but also to bend the rules of reality—without dumping everything on a reader at once—a bit like quantum physics. Like an onion, each arrives in layers as the pages of philosophy, ancient wisdom, spirituality, and science are peeled away—but once we arrive at the center, what is left?

Additionally, Sci-Fi writers should be comfortable with complexities that involve blood, technology, mystery, politics, clandestine operatives, complex systems, and AI—and concisely conveying such heady topics in an entertaining and relevant way—one that provokes new lines of thought, appeals to the human condition, or potentially enriches life with relevancy.

The balancing act of composing decent Sci-Fi flows between the logical and the lofty—a grounding to Earth and an ascension to portals of creativity where the unimaginable possesses weight and soul— a bit like science and the sacred. So, this also requires tolerance for uncertainty, ambiguity, ethics, and exercising empathy and non-judgment across disciplines and timelines. Like quantum physics, great SCI-FI raises questions that might not ever be answered.

As we consider Sci-Fi as it might pertain to DNA, some semifinal thoughts on the subject:

- DNA could be used as storage and retrieval for on-demand ancestral memories or consciousness, a bit like firing up a computer and searching through files
- Genetic manipulation could be summoned with a ritual, electricity, or sensory deprivation to enhance, bring forth, or transfer consciousness to a screen
- DNA-based quantum computing could lead to artificial or hybrid biotechnological consciousness (this may no longer be in the realm of sci-fi, but a reality)
- DNA might be harnessed as a receiver/transmitter for universal consciousness, capable of hitching a ride on neutrinos for instantaneous information transfer

A Spiritual Perspective of DNA

DNA is often viewed in the spiritual context as a sacred blueprint or code, a manifestation of a higher power or elevated soul. It is said to connect individuals to their ancestors and/or the universal life force or consciousness. Someone once said that the soul of a person is like an information cloud, akin to, and part of, a highly intelligent computer.

Conversely, interpretations of spiritualism and consciousness vary across different cultures and traditions but typically include:

- **Divine Design**: complex code latticed with the potential for life and individual characteristics, reflecting a higher creator's plan for each being.

- **Karmic Imprints**: DNA carries karmic imprints from past lives, influencing present traits and experiences and highlights the interconnectedness of all beings.

- **Soul Blueprints**: DNA is a physical manifestation of the soul, storing information about our ancestors, epigenetic memories, our ongoing spiritual journey and our potential.

- **Collective Consciousness**: DNA is individualized yet also linked to a larger consciousness collective, a co-op, which engages individuals to connect, exchange energy, and share experiences.

- **Spiritual Evolution**: Through conscious development and intention, individuals can and do influence their DNA to evolve physically and spiritually. (Unfortunately, this development and intention set in reverse can have the opposite effect, causing regression or other devolving sufferings.)

Ancient Wisdom Traditions (Non-inclusive)

Ancient wisdom traditions—like those from Vedic India, Indigenous cultures, Taoism, and Hermeticism—suggest that consciousness is fundamental to life and intimately connected to the fabric of the universe, including the human body and its DNA.

- **Hinduism**: "Atma" (soul) is linked to DNA, where the soul's essence is believed to be carried within our genetic makeup. Hindus often greet the day by finding holy spirit in *everything*, including other people and religions. They seem to adopt an enlightened philosophy that anything available to our senses as matter, originated from the ether as an idea—and this in and of itself, even if its mud or a pile of garbage, is sacred—because matter has been made manifest from nothingness.

- **Buddhism**: Interconnectedness, reincarnation, and karma align with the view of DNA as a threaded code and ladder connecting all life, carrying the consequences of past actions.

- **Native American Traditions**: DNA has an unbreakable connection to the land, nature, and ancestral spirits (especially the sky gods), emphasizing the importance of respecting our environment and planet.

- **Sufiism**: Love is central in Sufism—and its *core engine*. God is Love itself. The human soul eternally yearns for union with the Divine. This is expressed in passionate poetry by mystics like Rumi, Hafiz, and Rabia, which espouse love, service, spiritual refinement, and inner transformation. Also central to Sufism is the idea of Oneness—an interconnection of everything, similar to Bohm's *hidden variables*. According to one Sufi text: "*We created man and breathed into him of Our spirit.*" Who is *we*?

Rare Bloodlines and Rh-negative Significance

In 1937, Karl Landsteiner and Alexander Weiner discovered a new blood type: the rhesus blood type, or Rh factor. They were researching solutions for the cause of a medical mystery that killed dozens of babies each day.

Rh- (Rhesus negative) refers to the absence of the Rhesus (Rh) factor, a protein found on the surface of red blood cells. The Rh factor of one's blood is an inherited trait. Almost ninety-percent of people on this planet have the Rh+(positive) blood protein, while those with Rh- negative blood do not. Blood types are categorized into four main groups (A, B, AB, O), and each can be either Rh+ or Rh-, creating eight total blood type categories.

Globally, and presently, only about ten percent of the entire global population has Rh- blood and the prevalence varies by ethnicity, particularly certain isolated populations like the Basque people, found in a mountainous area between Spain and France. The Basques have an unusually high frequency of Rh- blood, at almost forty-percent, raising intriguing questions about its origins. Caucasian and Indian populations also tend to have the highest percentages of Rh- blood types. No one knows why.

Other rare blood is known as *Golden Blood* (Rh-Null). Fewer than a dozen people in the world have Rh-null blood as they possess zero Rh antigens on their red blood cells. The first case of golden blood was discovered in an Australian aboriginal woman, while other humans boasting this blood come from a variety of ethnic backgrounds. It is a mutation of the RHAG gene, but no one seems to know its origins either.

Moreover, the genetic variations of Rh- and Rh-null are older than humans are as a species. Additionally, Rh- is a recessive trait. A recessive trait is only visible when an individual inherits it from *both* parents. Additionally, There are now only about ten Rh-null or *golden bloods* left on the planet.

One wildly perplexing thing about Rh- blood is the fact that if an Rh- mother carries an Rh+ baby, her immune system will recognize the baby's Rh+ red blood cells as foreign, and start producing antibodies against the developing fetus, as if it were a foreign body! This leads to something called hemolytic disease, where a baby might be born deaf and/or blind—or worse.

This condition is medically referred to as *Rh incompatibility*. It is managed with an injection of Rh immunoglobulin (RhoGAM, developed in 1968) administered within seventy-two hours of delivery to prevent subsequent fetuses from being attacked by the mother's antibodies.

Intriguingly, Rh- mothers and other Rh- people are recorded on governmental registries. Rh- mothers are also issued cards or other means of ID to inform medical personnel of this rare blood type. An Rh- person may only receive a blood transfusion from another Rh- person—but an Rh- blood type can donate blood to anyone, making them universal donors—which feels a bit counter-intuitive.

During my first pregnancy, I was told at the doctor's office that I needed to give a blood sample. I lived in a rural village with only one obstetrician. When my results came back, the doctor was practically giddy with excitement and shared that in his thirty-year career, he'd never had a mother present with Rh- blood. I'm still unsure of why this animated him so much, but I had to sign a federal form and was given a card to carry. I was nineteen, and impressionable—so I signed the document, took the card, and went on my way, later delivering a healthy almost nine-pound baby boy.

Similarly, some studies suggest that Rh- individuals have unique immune responses compared to Rh+ people. During the Covid-19 pandemic, Rh- blood types had between a twelve and twenty percent *lower* chance for developing full-blown CV-19. They are also shown to be less susceptible to AIDS, human papillomavirus (HPV), or malaria. Origins of the Rh- factor remain unknown. This has prompted speculation about genetic mutations, adaptive traits, or even *non-human ancestry* (as in extraterrestrial) hypotheses.

Furthermore, there have been rumors suggesting that intelligence agencies and laboratories have conducted classified research into Rh-individuals, particularly for their perceived uniqueness. Rh- blood types seem to be more open to experience, fall into trance states more easily, and tend to make effective remote viewers, although much of this is speculative and not (as far as we're told) scientifically tested.

The CIA's Stargate Project reportedly included participants with diverse genetic and neurological traits. However, there is no official evidence that has ever surfaced that Rh- individuals were specifically targeted or studied within Stargate.

Some declassified biological and medical studies have examined how genetic traits (including blood types) influence immunity, disease resistance, and psychological traits, but these studies have been general rather than specific to Rh- individuals.

Likewise, Rh- blood has been linked to speculative theories within ancient astronaut hypotheses or connections to lost civilizations. While these ideas completely lack empirical support, they have fueled public intrigue, great fiction, copious conspiracy theories, and cultural narratives.

From a quantum physics perspective, let's take a small leap into how Rh-negative blood types might offer individuals quicker and more expansive access to other dimensions, higher states of consciousness, or perceptive abilities.

We might hypothesize that the Rh protein's absence in Rh- blood might create unique quantum properties at the molecular level. Perhaps the missing protein creates "quantum gaps" in the cell membrane structure that theoretically interact with quantum phenomena in unusual ways.

We might imagine that these quantum gaps could make Rh-blood cells more susceptible to quantum entanglement effects. In this scenario, the absence of the Rh protein and its respective spikes might allow blood cells to maintain quantum coherence for longer periods than typically possible in biological systems, like how some scientists study quantum effects in photosynthesis.

Taking this further, we might envision that this unique quantum behavior might explain some of the folklore and mysteries that have historically surrounded Rh- blood types. For instance, if Rh- blood cells do maintain quantum coherence, they might theoretically be capable of quantum information storage too, where memories or information could be encoded in quantum states within the blood cells—which seems to somehow happen with both blood and DNA in the CIA-encouraged Baxter and later, the Marina research, even among most Rh+ individuals.

It's been postulated that due to enhanced sensitivity to quantum fields, which most other humans can't detect or only cursorily perceive, Rh- individuals are able to form quantum networks and even communicate with other Rh- individuals at great distances telepathically—like how some autistic children reportedly meet at a parallel place known as *The Hill*.

The *Telepathy Tapes* is a podcast series hosted by Ky Dickens, a documentary filmmaker. She explores the idea that some non-verbal people diagnosed with autism possess telepathic capabilities. She interviews parents, teachers, counselors and neuropsychologists, who've mentioned remarkable, mind-bending experiences while engaging with primarily non-verbal autistic children. Allegedly, these children (and some *invited others,* such as a beloved teacher or parent) meet in a telepathic parallel universe, a sort of internalized 'chatroom' called *The Hill.*

While mainstream science takes issue with what it deems pseudoscience, Dickens offers a refreshing perspective that is backed up with research—and not just hers. I ask, if scientists have developed machines capable of wirelessly interfacing with the human brain—enabling communication for individuals with conditions like ALS or mind-maneuvering drones—why not explore the potential for mind-to-mind interfacing? History shows that what we often dismiss as pseudoscience or the supernatural frequently turns out to be science or energy we simply don't yet understand.

Alternatively, scientists do not know how or where Rh- blood originated and, by being older than the human species, what does this mean?

We continue in the realm of science fiction, based in at least a little bit of science, which determines, so far, that if Rh-negative blood is *basic* blood—meaning it could be quite primitive rather than magical or other-worldly, what changed in the evolutionary cycle that added spikes to blood proteins of ninety percent of the global population but left ten percent with smooth, sometimes slightly egg-shaped red blood cells?

We either have a missing link(s) in evolution that is so primal it has gone underground and been lost —or so "out there" it may have extraterrestrial origins. You can bet only the bravest and most imaginative of scientists would dare take on such a sci-fi sort of experimental challenge. Yet it begs the question: why would nature (or God), engineer or enable such a biological anomaly that normal human physiology doesn't support?

The Brain-Matter Interface

The brain-matter interface (BMI) is technology that enables direct communication between the brain and external devices by recording neural activity through implanted electrodes or non-invasive methods, translating brain signals into digital commands that control computers, prosthetics, or other devices, and then sends sends information back to the brain through electrical stimulation.

A couple of examples include Neuralink's brain implant chips, Synchron's stentrode device, and Blackrock Neurotech's array systems. Clinical applications focus on restoring movement and communication for people with paralysis or neurological conditions. Challenges include biocompatibility, longevity of implants, signal processing accuracy, and ethical considerations around privacy and autonomy.

Such neural interfaces are also open to ethical concerns and potential for abuse as they become ever more sophisticated. And—like a Sci-Fi epic, advanced technologies do exist that interpret brain signals (EEG-type headsets), or other subtle behaviors exhibited through phone usage to infer someone's thoughts or emotions or read retinal impressions—at least up to a rather frightening degree of certainty.

Alternatively, other technology exists where microscopic integrated circuits are implanted under the skin, commonly the gap between the thumb and forefinger, for a range of clandestine purposes. Universal scanners typically set at particular frequencies are extraordinarily effective at collecting information or tracking a biological source.

Moreover, it might be possible to use electro-chemical processes to power implantable generators within our bodies. The average human body expends about the equivalent of eight-hundred AA batteries of energy per day, yet its mechanical efficiency hovers around an average of twenty-three percent. Unfortunately, most of the energy we gain from food is released as heat.

Presently and unfortunately, most current cardiac pacemakers, implantable defibrillators, neurostimulators, drug delivery systems, and bone growth generators mostly rely on lithium-ion batteries, which must be surgically replaced as they age.

To solve some of the issues with implantation devices, researchers continue to develop implantable *generators* that harvest energy from human bodies to power a new generation of devices known as TEG's (thermos-electric generators).

Alternatively, a man named Roger Leir (DPM), who died in 2014, claimed to have surgically removed from his patients, strange implants that emitted radio waves, had magnetic properties, or possessed crystalline structures.

Most of Leir's patients also claimed to have had alien abduction experiences. Notably, many of these objects were extracted from the hands and feet of his patients, lending skeptics a highly reasonable opportunity to point out that these *so-called* alien implants were nothing but ordinary glass or metal fragments that lodged in a body while walking or due to some other hands-on activity.

Leir countered in an article just before his death that the implants he removed from patients had undergone rigorous examinations by both biological and metallurgical scientists associated with some of the world's most prestigious laboratories, including but not limited to, Los Alamos National Labs and Seal Laboratories. The discoveries, he said, proved beyond the shadow of a doubt that these objects are highly suspect in origin because no evidence was found for inflammatory or rejection reaction by the body and there was no visible portal of entry from the objects.

Moreover, the existence of Carbon Nano Tubes, Carbon Nano Strands, and Carbon Nano Fibers were discovered in some of the implants, something once considered by mainstream scientists *nonexistent in nature* and nonreplicable in a lab—at least until they reversed course when commercial products were later designed and made from this very same technology—and is today, considered some of the strongest substances known to humankind.

Either way, my first question as a researcher and a Sci-Fi author is, "Why?" Then, was it *aliens* or was it "us" (as in covert military or secret intelligence)?

PART III

Quantum Controversies & Cutting-Edge Physics

CHAPTER NINE

The Mind-Matter Interface, Consciousness, and Information Theory

The mind-matter interface concerns the philosophical relationship between consciousness/subjective experience and physical reality. This raises fundamental questions about how subjective experience emerges from or relates to physical processes, and what is called the "hard problem of consciousness."

A "mind-matter interface" is a concept where the human mind directly influences physical matter—suggesting a connection between conscious thought and the physical world. While this interface is largely relegated to parapsychology, it basically implies that mental intention affects events even at a distance without known physical mechanisms.

Here's a wild experiment that seems to back up the mind-matter interface: Nikolai Kozyrev, a Soviet-era astrophysicist believed that time is more fluid than we assume (more on time in a later chapter). He hypothesized that time wasn't some passive dimension but a force interacting with matter—and consciousness. He used mirrors to induce altered states of consciousness in participants. The mirrors were alleged to create a unique enclosed space that altered *perception* and blocked out electromagnetic influences.

Participants stepped into the chamber of mirrors and reported telepathic communication, astral travel, out-of-body experiences etc. Yet the most impressive part was when scientists expanded the experiment around the globe to some 5,000 subjects and then transmitted information to them while in the chamber. In a few cases, 95% of what was received was correct.

It opened the idea that human consciousness sometimes operates nonlocally without spacetime restrictions. UFOs were also reported above mirror chamber research centers, as if the mirrors served as some sort of beacon or portal to another dimension. (Makes me rethink those neutrino detectors listed in an earlier chapter.) Of course, Kozyrev was skewered by the scientific community and his work relegated to the field of pseudoscience—at least publicly.

So, while the brain-matter interface is primarily studied through neuroscience and bioengineering, the mind-matter interface involves deeper philosophical questions about consciousness, qualia, and the nature of subjective experience that goes beyond purely physical descriptions. If we return to the placebo effect (and we will), which has been repeatedly documented and replicated as real phenomena, wouldn't this also be an example of mind-matter interfacings?

Additionally, it is a concept related to John von Neumann's interpretation of the quantum measurement problem, and Schrödinger's cat experiment, mentioned back in chapter two, namely that consciousness causes the collapse of the wave function when a quantum system in a superposition of states is observed. This magical cat has served as basecamp for many theories, including Bohm's idea of hidden variables, the implicate and explicate orders, holomovement, and quantum potential.

Since consensus over the quantum measurement problem has not yet been reached, further attempts to provide empirical evidence on mind-matter interface should be widely encouraged. In fact, mind-controlled or *brain drones,* as they're sometimes called, have been invented and flown or operated with success since the early 2000's. If we can wirelessly link man (or woman) to machine to communicate, should we not also be able to transmit telepathically person-to-person, especially if human beings are nothing but slabs of meat and fat, processing information through a brain?

Later, in 2017, university labs and private citizens also successfully built the same types of drones. If we can fly drones using our minds, we can certainly scale it up to control weird craft designed to scare the shit out of unsuspecting citizens—but for what purpose other than an uninformed consent psyops experiment, remains unclear.

Complications of Consciousness

But what, exactly, is consciousness?

Consciousness, like quantum physics, has no single agreed-upon, provable place in the brain or even a concrete definition in science—but like dark energy and matter, we all talk of it as if it exists. So, like a sound wave—is it an as-of-yet undiscovered exotic matter? Or is it maybe a photonic grouping of neutrinos? Descartes *dualism* contends that consciousness is separate from physical matter, suggesting a split distinction between mind and brain.

Panpsychism, on the other hand, proposes consciousness is a fundamental feature of reality existing in even the simplest of particles. But if particles don't exist the way we think they do and everything is but one field of flowing and folding existence, what might this mean?

Now we arrive at energy-consciousness where some theories link consciousness to the universe's energy fields, suggesting awareness arises from or interacts with these fields—that all beings have an underlying essence much the way quantum physics has some sort of substrate beyond subatomic particles.

For example, biologist Rupert Sheldrake suggests that all living organisms are guided in their development by invisible "fields" that carry information about form and structure. This morphogenetic field essentially acts as a blueprint for an organism's development. He also alleges that these fields are influenced by the collective experience and consciousness of previous similar organisms through a process called "morphic resonance", where a developing organism "resonates" with the morphogenetic fields of similar organisms (think schools of fish, flocks of birds, or crowds of humans all flowing synchronistical).

Morphic fields are considered non-physical entities, meaning they too, like most quantum physics and consciousness, cannot be directly measured by conventional scientific methods. This tells me that it is time for science to rethink its paradigm into something that transcends the empirical model, which, both frustratingly and fortunately, is due in great part to the *quantum* of things.

Critics argue that morphogenetic theory is difficult to test scientifically as it is not clearly defined and is adaptable to explain almost any phenomenon—like most religions, omens, or superstitions. Yet there are scientists revisiting this theory and taking a closer look.

One experiment that appears to provide backup to Sheldrake's theory is the Global Consciousness Project (GCP) begun in 1998 at the Institute of Noetic Sciences, founded by Dr. Mitchell after his trip to the moon and back. It aims to study whether collective human consciousness, particularly during significant global events, influences the output of random number generators.

The project strongly suggests a link between human mind energy or universal consciousness and physical systems; essentially, our collective attention affects the physical world in a measurable way. The project's researchers maintain that their data shows significant correlations between major global events and deviations in random number generator outputs, suggesting support for the concept of a collective consciousness.

"*Not so fast!*" skeptics say. Critics severely question GCP's methodology and claim that observed anomalies are likely due to selection bias and "pattern matching" rather than evidence of a global consciousness. Replication experiments have also seemed to fail. Alternatively, proponents claim that some aspects of the experiment result in downgraded results due to participant boredom, disbelief, and tedium.

Notably, the GCP is an extrapolation of two decades of experiments from the Princeton Engineering Anomalies Research Lab (PEAR), which was also publicly skewered and considered an embarrassment by establishment academia for similar reasons. It closed its doors in 2007.

Today, the "Global Consciousness Project 2.0" is managed by the HeartMath Institute. Doc Lew Childre Jr. is its founder. He is the son of Doc Childre Sr., the Grand Ole Opry star known for the song, "*Let's Go Fishing*".

The HeartMath Institute conducted experiments in how sounds align (or misalign) a person's heart and mind energies; for example, by monitoring EKG and blood pressure of subjects while they listen to various sounds, those frequencies could be tailored to invoke a specific response, like tuning fork work...or music. We all have songs that make us want to dance, crawl into a corner and weep, lift us up, or fill us with aversion and dread.

Incredibly, music was used as a device of torture for detainees held by the United States during the War on Terror. Interrogates opted for a combination of heavy metal, rap, country, or children's music, typically played on repeat for hours. Music as an instrument of torture originated in psychological research from the 1950s, and the tactic mixed in physical discomforts like stress positions or severe room temperature fluctuations. It was officially approved by several prominent military officials throughout the globe.

Musical tuning had its grounding much earlier, preceding World War II, when there was no universal tuning standard. Musical standards varied wildly between 400–460 Hz depending on country, city, and composer. For example, baroque music often used a lower tuning (~415 Hz), while some romantic-era orchestras tuned higher (~450 Hz). In 1936, the Reichsmusikkammer (Nazi State Music Bureau) adopted 440 Hz as the official tuning frequency.

Joseph Goebbels, the Minister of Propaganda for the Nazi Party, allegedly later supported its use for propaganda purposes. His action was cited as the beginning of "weaponized tuning," but there's no direct evidence the Nazis chose 440 Hz over 432Hz to intentionally manipulate consciousness.

The International Standards Organization (ISO) adopted 440 Hz in 1955 (as ISO 16), but the pitch had already gained traction after being endorsed at a 1939 conference in London. It was largely promoted for practical reasons, like standardizing instrument manufacture and ensuring uniformity across global performances. The U.S. adopted 440hz early, with heavy lobbying support from J.C.

Deagan, Inc., a tuning fork manufacturer—who had zero affiliation with Nazis.

Notably, tuning forks are simple tools that vibrate to produce specific resonant frequencies when struck. They were invented by John Shore in 1711 and are commonly used in scientific experiments, for tuning musical instruments, and even in medical and therapeutic settings to detect nerve damage or hearing loss.

Interestingly, both the CIA and the military further conducted research to evaluate the success rates of *test-bed* and *field trials*, based on PEAR's Global Consciousness model—and sometimes used frequencies to enhance faster trance states.

Remote viewers were asked to sense and describe either natural scenes or military sites featuring designated agents or beacons. Both the 'transmitting agent' at the target site and the 'viewer' or 'receiver' sitting in a lab was asked to fill out an identical thirty-point question-naire with a yes (1) or no (0) marking. Random number generators also used 0's and 1's.

The CIA and military assessment methods were first developed by Princeton University researchers at the Engineering Anomalies Research Lab while investigating 'Precognitive Remote Perception'.

Using advanced mathematical methods developed in the field of artificial intelligence and pattern recognition, the degree of success of the remote viewer was quantified, as we'll see, way beyond chance. Every experiment, on some level, even if unrealized at the time, appeared to measure (likely without recording), perception and intention.

Ultimately, the complicated question, ***"What is consciousness?"*** remains central to all the 'deep-in-the-matrix' ongoing clandestine research (and rituals) in science, religion, philosophy, spiritualism, and quantum physics. But three scientific approaches do attempt to describe consciousness:

- **Global Workspace Theory**: Conscious experiences are information broadly available to different brain systems
- **Integrated Information Theory**: Consciousness emerges from information being integrated across neural networks
- **Higher-Order Theory**: Consciousness requires *meta-cognitive* awareness of mental states. Metacognitive awareness is the ability to personally and internally recognize how one thinks and learns, and to use strategies to monitor and regulate learning. It involves stepping back from thoughts and emotions, observing them as passing mental events. In spiritual practices, we'd likely call this *mindfulness*.

The challenge humans have with consciousness lies in connecting objective neural activity to subjective experience, which neuroscience can observe but not fully explain mechanistically or specifically locate in a human brain.

The *hard problem of consciousness* first coined by philosopher David Chalmers in 1995, refers to explaining how and why physical processes in the brain give rise to subjective, conscious experiences—the *feeling* and *experience* of what it's like to see red, to feel pain, or to have thoughts. While science explains neural mechanisms behind perception and behavior (the "easy problems"), accounting for subjective experience itself (qualia) remains out of reach.

A scientific theory bridging the gap between objective brain activity and subjective conscious experience doesn't yet exist—a bit like the substrate space or dark energy that separates classical and quantum physics.

Maybe this means that consciousness is a hotbed of neutrinos or… like a neutrino, moves through everything, and gives rise to experience rather than being harnessed for precise measurement.

The Penrose-Hameroff theory at least represents an attempt to meet quantum physics with human consciousness, and though it remains suspect in the empirical realm of science, it has great merit, and much of it "feels right"—but feelings are not science.

Coincidentally, suppose we try, like Einstein did, a *thought* experiment—except ours is to consider how placebos and *belief* might interact as a bridge between classical and quantum physics.

What if quantum entanglement between particles in neural networks facilitates faster-than-classical information processing, potentially allowing belief states to rapidly influence biological systems? This assumes, of course, that thought moves faster than the speed of light.

And here's where shit gets crazy. According to Einstein's special relativity, nothing with mass—including thoughts, which allegedly arise from physical processes in the brain, can travel faster than light speed (approximately 3×10^8 meters per second). While it's true that neural signals travel relatively slowly, around 120 meters per second, through chemical and electrical processes—a thought, we're told, does not possess mass or matter.

Thought itself is considered an emergent *phenomenon*, like information, or a pattern of activity. This sounds a lot like quantum physics to me! Furthermore, the measurable energy consumed during thinking is extremely small and the brain uses about 20 watts of power total, with individual thoughts using only a tiny fraction of that.

Through $E=mc^2$, this energy produced by thinking *technically* has an equivalent mass, but it's vanishingly small and not meaningful to measure at the level of individual thoughts—a bit like neutrinos.

Yet I'm reminded that despite both being challenging to detect and measure, thoughts and neutrinos operate through fundamentally different mechanisms—unless Penrose and Hameroff are onto something—and we've all been forced off a hot trail by being steered to a more widely lukewarm and accepted one.

Using the quantum Zeno effect, where observation affects quantum states, might provide a mechanism for consciousness (including beliefs) to influence quantum-level biological processes.

However, while quantum effects in biology are increasingly documented (like in photosynthesis and bird navigation), direct evidence linking *belief* to quantum processes remains speculative despite available evidence of the placebo effect. We cannot measure or detect universal consciousness, explain how individual awareness emerges from a universal field, or understand the mechanism connecting quantum effects to conscious experience—but it's hidden right here on Earth in plain sight, within the nocebo and placebo!

Then we have *information theory,* which claims everything is a field of information, especially if applied to the universe at large. Prominent figures like physicist John Archibald Wheeler advocated for the idea that "it from bit" reality is fundamentally composed of bits of information rather than matter itself. Someone beat him to that though many years earlier, albeit in a different form.

Information theory helps bridge gaps between quantum mechanics and general relativity. It might even offer new insights into concepts like dark matter and dark energy—but I feel that even here, we are being subtlety steered back to the gravitational model and crammed into Schrödinger's box, along with the cat!

What might happen if we applied information theory to an electrical-plasmic universe?

It was American mathematician and computer scientist Claude Shannon who conceived of, and laid the foundations for, information theory during the 1940's (*The Mathematical Theory of Communication*, Shannon and Weaver, 1949 University of Illinois Press).

Working at Bell Labs, Shannon's theories also laid the groundwork for electronic communications networks that currently web Earth. He thought about gravity sometimes—but he rarely relied upon it for the basis of his work. Again, it was what he didn't say—and what was left out of his publicly available mathematical equations, that spoke volumes.

As a mentor to Vannevar Bush (who, like Oppenheimer, was also a leader on the Manhattan Project), Shannon once said that it is only a small percentage of the human population, which produces the highest proportion of important ideas. He also believed that great insight sprang from a vying combination of curiosity and dissatisfaction (much like this writer). And he liked to play. He was, for some odd reason, fascinated by juggling. Curiously, he owned an electronic mouse named Theseus (a divine hero in Greek mythology who slayed a minotaur).

Shannon cracked the universe's code of abstract information, reduced the noise around it, and broke it down into bits. He realized that those bits of information, or bytes, could be placed in circuitries and sent around the globe and into space nearly flawlessly. In other words, information from somewhere *outside* could be successfully placed *into* closed systems—as long as those closed systems remained insulated from environmental harm—maybe like brain microtubules.

Shannon proved that all matter of gear and gadgets, electrically zapped, could process information (he once made a morse code machine out of barbed wire similar to the way Tesla placed lightbulbs in dirt and used *Ether* to light them up without wires).

Shannon also knew that the electron (an alleged flavor of neutrino) moves through wires and that things that seemed to have nothing in common had everything essential in common. He too, knew that empty space is not really empty—but pervaded by a nonzero expectation value, even in the vacuum of space—which he viewed as a sort of superconductor (reminding me of Tesla's thoughts). He also focused on the *structure* rather than the *meaning* of information, as a mathematical abstraction, and viewed information as a way to reduce *uncertainty* (entropy) in signal transmission.

Shannon laid it out for everyone to see way back in the 1940's— but hardly anyone knows his name. He gifted us the sight to *hidden variables* that Bohm had originally and so beautifully unveiled. But Shannon warned of a darker side to such quantum manipulation: He firmly believed that humans would one day create fully bio-conscious and far superior artificial intelligences—and that this AI would someday outsmart, conquer, and completely rule over humans until the moment of complete extinction of homo sapiens from the planet. Note, a *conscious* artificial intelligence is not the same as having *consciousness*. A strictly logical machine that can feign empathy to meet an algorithmic goal has zero capability for compassion. If you're in its way, you're removed.

Consciousness requires an active agent in the field, not just an observer or transmitter of information. It needs a *participator* and is foundational to *reality* rather than emergent. While information can be copied and moved, it has no *meaning* in a sentient sense. A machine can replicate you or claim to represent you but can never fully reproduce *you*. Even a biological clone is not *you*.

It is our attention and awareness that turns potential into reality. Experiences in the invisible state cannot be reduced to numbers in complex biological systems the way information is fed into computers, quantum or otherwise. In other words, a computer is a map of information while a human is a quantum field possessed of meaning. Algorithms simulate structure, not meaning, and a machine cannot and will not ever understand "being".

Like da Vinci, Shannon was a thinker and a tinkerer. As an engineer, a mathematician, an inventor, a chess master, and a juggler (among many other things), he defied specialization and categorization. And like da Vinci, he roamed wherever curiosity took him.

Both men found the links to unlock quantum codes using *intuition* and their minds at play highly benefited science, technology, and humanity. It is safe to say that both da Vinci and Shannon were creators beyond pure logic. We are all creators, which means that the universe and God (however defined) is not separate from us.

But regarding Artificial Intelligence overtaking humans—I pray to God that Shannon was wrong.

CHAPTER TEN

Plasma Consciousness, Orbs, & Cold Fusion? Oh, come on!

M ainstream publicly available science has ***never*** developed a theory that directly links plasma to consciousness. Still, discussions around plasma in the context of consciousness emerges in more speculative or interdisciplinary domains like panpsychism, consciousness studies, or astrobiology.

Furthermore, "plasma consciousness" isn't a widely established term in either scientific or philosophical discourse. However, from a science and science fiction perspective, plasma exhibits complex, self-organizing behavior and responds to environmental changes, potentially giving rise to some form of consciousness.

If plasma merges with biological entities that exhibit awareness, humans might mentally control spacecraft or other sorts of technology like, oh, say, time machines, weaponry, and drones.

The most common mention of plasma consciousness in science circles appears briefly in discussions about ball lightning and other atmospheric plasma phenomena, with some speculating these could exhibit very primitive forms of awareness, meaning plasma lacks a central nervous system or self-awareness, but might respond to an environment in a seemingly coordinated way.

As previously examined, plasma is one of the four fundamental states of matter (along with solid, liquid, and gas) and prolific in our universe. It consists of ionized particles, including positively charged ions and free electrons, and is found in stars, lightning, and certain high-energy environments. And it is a far better conductor of electricity than even copper.

Alternatively, consciousness refers to the state of being aware of and able to think and perceive. It involves subjective experiences, thoughts, emotions, and sensory inputs.

Admittedly, the idea of "plasma consciousness" is highly suspect in a strictly material world but if we merged plasma with 'consciousness', we might explore how consciousness arises from or is connected to plasma-like energy or its fields, suggesting that the behavior of plasma might potentially mirror or aid conscious awareness in a non-biological way:

- **Analogies between plasma and consciousness**: Plasma is dynamic and exhibits complex, often unpredictable behavior, which might serve as a metaphor for the dynamic, fluid nature of consciousness, through fluctuating states that could be likened to thoughts or emotional states.

- **Consciousness in plasma-based lifeforms**: In speculative science fiction, imagination abounds about intelligent life forms existing within or using plasma states or environments, which might give rise to a form of "plasma consciousness" or plasma being—like an angel or a ghost.

It's been reported in various media outlets that both American and Russian astronauts have seen 'light' beings with wings in space, allegedly peeking into craft windows. However, there's been zero concrete evidence or confirmation for this. The reports I read online are likely based in conjecture and/or propaganda—or advanced secret technology designed to *freak out* the Russian astronauts and/or other respective governments—perhaps a highly technical prank. Still, others whose feet have never left Earth have reported seeing such beings, so the jury, at least for me, is still out.

Additionally, the idea of harnessing plasma to display conscious-like artificial intelligence (AI) is a fascinating concept that has ventured into the realm of advanced theoretical ideas. Did she just say, *ADVANCED THEORETICAL IDEAS?* While some of us easily dismiss *speculative* science let's explore the *advanced* part.

Plasma is an ionized, electrically conductive state of matter, where electrons are free from atoms. It behaves in ways that are vastly different from solid, liquid, or gas states, exhibiting complex and often chaotic dynamics. In some ways, plasma might be considered a "smart" material. The key properties of plasma—such as electromagnetic interactions, self-organization, and sensitivity to external forces—are often studied in fields like plasma physics, but they are typically not associated with conscious awareness.

Consciousness, as we already read, is a highly complex foundational phenomenon typically linked to biological systems, particularly brains. While AI systems *simulate* certain cognitive functions (like pattern recognition, learning, and decision-making), they don't possess subjective experience or soul, which is considered a core aspect of consciousness.

Consciousness involves self-awareness, perception, emotions, and the ability to have experiences, which are difficult to recreate through purely computational processes outside the biological context. Feelings, intuition and experience are subjective and cannot be precisely measured, dominated or monetized because each is impermanent.

While plasma is complex, it is not considered a material that inherently exhibits properties of consciousness. Consciousness in humans and animals, arises in part, from interactions and organization of neural networks in the brain, a highly specialized biological structure, which appears at face-value to be computer-like, but is highly likely an entangled quantum state or field. While a biological body or mind-brain can be predictable, manipulated, or controlled, consciousness and its awareness remain free and dynamic.

Plasma, as we learned, is a physical state. While it could potentially support certain advanced technologies (such as plasma computers or plasma-based processing systems), it is unclear how these systems might lead to conscious experience—unless we somehow combined and uploaded it with organic biological material, such as replicating neural networks derived from human brains. Maybe this could be done in a lab setting using petri-dishes, where the networks are then transferred and encouraged to replicate inside a quantum computer—but again—would it have consciousness or simply be conscious?

Alternatively, this hotly debated topic provides little concrete scientific evidence to suggest that plasma, on its own, exhibits or "harnesses" consciousness. On the other hand, plasma has been used in certain covert experiments to simulate certain aspects of intelligence, but whether it could independently generate subjective experience like human consciousness remains uncertain. If it were so, we can speculate that such information would be withheld and classified for the fear, societal breakdown, and panic it would incite.

In theory, if artificial intelligence were developed to the point where it could simulate or approximate consciousness, it could potentially be housed within a plasma-based computing system (perhaps how quantum computing might be used to process complex AI algorithms).

Plasma-based systems offer unique advantages, such as ultra-fast information processing or the ability to store vast amounts of data, but again, this wouldn't necessarily equate to consciousness—it would just be an advanced form of AI.

If plasma were used in the context of AI, imagine a highly advanced system where plasma's electromagnetic properties help simulate cognitive functions on a massive scale, such as mimicking the behavior of neural networks or brain-like processes.

In the far reaches of science fiction, one also wonders if a sufficiently complex plasma system could "generate" consciousness, in ways akin to the way the brain organizes its neural pathways. This requires breakthroughs in understanding both the nature of consciousness and the physical properties of plasma. The public is not yet there—and scientists who've had such breakthroughs often get their materials and research confiscated in the name of national security.

To briefly revisit, some scientists claim consciousness is strictly a biological byproduct of the brain—but like dark matter and dark energy, no specific location of consciousness in any brain has ever been found. Others think consciousness is a foundational and fundamental part of the universe, where human brains act as transponders, able to receive and send information because we are immersed in universal consciousness. This makes it difficult to design a system (whether biological, mechanical, or plasma based) that simulates or displays consciousness.

Yet hidden within the current scientific framework, consciousness is closely associated with highly complex biological systems (like the human brain) and with states of matter such as plasma.

Dr. Hannes Alfven was a pioneer in experimental investigations of plasma. He also referred to electricity as that one great branch of physics gone ignored in plasma studies. He began his career as an electrical engineer and developed models for understanding plasma as magnetic fluid.

In 1970, Alfven received a Nobel Prize for his discoveries in 'magnetohydrodynamics'. What hardly anyone knows is that he pleaded with his peers to ignore his early concept of magnetic fields 'frozen in' to 'superconducting' plasma. He felt that it was **wrong**. But because his work in this field set the foundation for accepted interpretations of magnetism in space, and paired wonderfully with the gravitational model of physics, it became scientific dogma—and scientists were reluctant to let it go. During his Nobel speech he warned that magnetic fields are only *one* component of plasma science, and their electrical currents **should not** be overlooked.

Furthermore, he felt there was a severe disconnect between mathematical theory of plasma and its lab experiments, which show that plasma absolutely refuses to obey its own math, no matter how elegant. Additionally, plasma cells moving relative to one another in space induce in each other, electrical currents—and can even form cell-like walls. This lends credence to plasma's ability to move through space as a conductor, developing electrical fields and magnetic fields as if in response to ambient magnetic fields resulting in complex behavior—but is it conscious? Most would say no.

Lacking training in circuitry or plasma discharge phenomena, I suspect that such experiments might render gravitational models practically obsolete. Why would electrical fields even acknowledge gravity? Gravity appears to be a weak, reluctant and passive-aggressive host for electricity, and easily overtaken by it—or perhaps a byproduct of electricity.

That gravity can generate enough force to separate ions from electrons seems *off*. It also throws the idea of black holes into the realm of shamanism, which might not be holes at all but dead stars holding exotic matter close to their chests—perhaps dark matter and energy? Or maybe something as simple as a glaze of photons expertly scattered and applied so thinly it just might make da Vinci weep.

Cold Plasma & Fusion

Cold plasma, also known as non-thermal plasma or non-equilibrium plasma, is a partially ionized gas where the electrons have a much higher temperature than the ions and neutral particles.

Unlike "hot" plasmas (like those found in the sun or fusion reactors), cold plasmas operate close to or at room temperature, making them suitable for a variety of practical applications such as making surfaces more adhesive or water-repellent, sterilization and disinfection in medicine and food processing, or air and water purification. The unique properties of cold plasma make it valuable for treating heat-sensitive materials like polymers, biological tissues, and electronics.

Cold plasma research continues to find new applications in fields ranging from agriculture to aerospace. But it leads me to question: Is it possible that our sun's outer photosphere is super-hot, and its corona a place of action, while the interior might be much cooler than suspected? Could this be why sunspots poke through and are attracted to one another enough to merge via magnetic polarity? Is the photosphere a sort of plasma shield around the sun?

Some of the most interesting material I've ever read regarding cold fusion and plasma comes from Dr. Takaaki Matsumoto at the Department of Nuclear Engineering, Hokkaido University, Sapporo, Japan.

Matsumoto discovered electro-nuclear collapse in a lab induced by an electromagnetic force (Electro-nuclear reaction (ENR), where completely broken materials regenerated again into tubes and filaments of carbon, oxygen and iron. The man and his team witnessed micro ball lightning that could completely break down and then rebuild materials—and in a way, it reminded me of our sun.

Plasma Orbs

Plasma orbs can be tuned to any frequency, any color of light, or they can be invisible, pulling electrons directly from the atmosphere. Creating lab plasma is relatively easy using a vacuum tube with two electrodes. Voltage is pumped into it and with the right tweaks, one can realize 'dark plasma'.

Dark plasma occurs through something called the Townsend regime, also known as Townsend discharge or Townsend avalanche. It is a physical process describing how electrical breakdown occurs in gases. It's named after John Sealy Townsend, who first discovered this phenomenon in the early 1900s.

For example, an initial electron is accelerated by an electric field between two electrodes in a gas. This electron gains enough energy to ionize gas molecules through collisions, creating more free electrons. The new electrons are also accelerated and create even more electrons through further collisions. The process creates positive ions too, which drift toward the cathode and release additional electrons through secondary emission. The Townsend regime is very important in plasma physics and the design of high-voltage equipment and gas lasers.

Consequently, plasma orbs can be created without emitting light. When plasma curves, glowing dots form. This is known as an abnormal glow discharge phase—but here's where it gets interesting—turning down the voltage creates a parallel state for **two** types of cold plasma—and the second one gives us cold fusion—equally distributed electrons and ions uniformly distributed. Such mass differences and charge create *free* energy. But I'd almost bet that at this point in vertical time, our best clandestine labs are already far beyond free energy.

The Correa Device

In the mid-1990s, a remarkable invention emerged from a laboratory in Ontario, Canada. Dr. Paulo Correa and Mrs. Alexandra Correa created something extraordinary: a device that appeared to generate more electrical power than it consumed, tapping into what they believed was Tesla's ether or the energy of space itself (Tesla also built a similar device and I found remnants of the design in a biography by Mark Seifer).

At the heart of their invention was a specially designed gas discharge tube, similar in basic construction to a fluorescent light but engineered for a completely different purpose. When operated under precise conditions, this tube produced powerful electrical pulses that delivered up to 900% more energy output than input and it was a result defying conventional physics.

The key to this cold fusion miracle lay in the tube's unique behavior. When voltage was applied, it created what's called an "abnormal glow discharge", a concentrated ball of ionized gas that exhibited bizarre properties. The Correas documented extensive experimental evidence showing these plasma formations could store enormous amounts of energy, like the mysterious "ball lightning" phenomena occasionally observed in nature.

The Correa experiments are compelling because of rigorous documentation. Unlike other alternative energy claims, their discovery was protected by multiple detailed U.S. patents that laid out precise experimental data, including careful measurements of input and output power using standard battery banks. The results were also clear and repeatable.

The author who wrote about the Correa device, Dr. Harold Aspden, connects the discovery to his own research and experiments suggesting that the vacuum of space contains vast amounts of accessible energy (voila Tesla!). He argues that the Correas had found a practical way to tap into this cosmic energy source, potentially opening the door to a new era of clean, abundant, and free power.

Yet most intriguingly, Aspden suggests that understanding the physics behind the Correa device could also help explain other mysteries, from the Earth's magnetic field to the formation of stars. The implications provided extended far beyond energy production, requiring a fundamental rethinking of basic physics.

The Correa device story is a lightning bolt of what might be possible if we're willing to question scientifically force-fed assumptions about the nature of energy and space itself. So, why didn't the Correa device, which seemed to uncover the mysteries behind the fundamental nature of energy and matter, ever achieve widespread adoption?

Like Bohm's *Hidden Variables* research, the device violated conventional laws of physics and the embedded understanding of thermodynamics, which claims that one can't get more energy out than one puts in (think of it like trying to get water out of an already empty bucket). Additionally, the mainstream physics community was deeply skeptical of any claims that contradicted these fundamental laws. Aspden notes that other scientists swiftly dismissed the Correa's findings "without even considering the evidence."

Moreover, while the Correas documented their work thoroughly in their patents, widespread independent laboratory replication of their results was not conducted. Scientists allowed scoffing, ego, or fear to stand in the way of rigorous empiricism, opening a clear path for unacknowledged special access projects to proliferate without eyes or knowledge. In this instance, it appears, the scientists dropped the ball and adopted the very same attitude they sometimes loathe in religious leaders.

The Correa's faced insurmountable institutional resistance from influential scientists who dared not court controversy at the risk of losing funding. Sounds a lot like Oppenheimer's instructions to established and up and coming physicists to severely distance themselves from Bohm's work on *hidden variables*—or JP Morgan's ostracization of Nikola Tesla and his free energy towers.

Granted, while the basic technology of the Correa device worked, there were legitimate engineering challenges to overcome, such as extending electrode lifetimes. The inability and the cost of this important hurdle pretty much wiped out any chance of large-scale commercialization.

Aspden's document doesn't directly address this, but historically, most alternative energy technologies face significant challenges from well-established, well-funded energy industries and black ops. These sorts of things are hoarded to maintain control and status quos. In a few cases, law enforcement and/or the military are weaponized to ensure inventors and dreamers remain within the crosswalk, obeying conventional science signals—or else. Now, if a researcher or scientist wants to 'voluntarily' hand over research to these groups post strong suggestions to do so, they receive a proverbial pat on the head and are told not to do that again—and like some farmers are paid not to grow crops—some savvy scientists, inventors, and entrepreneurs are subsidized to keep their research and designs on ice or in the dark.

Indeed, Matsumoto and his team in Japan explained cold fusion through the Nattoh model, where, rudimentarily stated here, hydrogen-cluster evolves many hydrogens—and it produces both black and white holes. He also warned that scientific thought about cold fusion needed to drastically change to understand such mechanisms. The important thing to note here is that as cold fusion progresses toward higher voltages and currents, it opens new fields for achieving maximized matter to energy conversion. It appears that many scientists will dismiss what they can't initially understand—and often—due to time constraints, refuse to learn from or hear out their peers.

Interestingly, Aspden predicted that acceptance of such technologies would likely come through practical demonstration to "the public at large, those who care little about where the energy comes from, so long as it is cheap, plentiful and non-polluting." Unfortunately, it hasn't happened and it's likely, it never will.

Ball Lightning

Nikola Tesla was fascinated by electrical phenomena, and ball lightning—a mysterious, luminous, and long-lasting electrical discharge—orbited within his field of curiosity. His interest in ball lightning aligned with his broader passion for understanding and controlling high-voltage and high-frequency electricity.

Tesla observed ball lightning phenomena during high-voltage experiments. He described glowing orbs of energy that appeared in his laboratory under certain conditions. Ball lightning had been reported by witnesses for centuries, but it is poorly understood by the public and still, to this day, lacks clear scientific understanding. Tesla sought to uncover the mechanisms behind it.

Tesla believed that understanding ball lightning would have practical applications for harnessing and manipulating electrical energy. He envisioned using such phenomena and plasma to develop advanced energy transmission technologies. His work often focused on high-frequency currents, plasma behavior, and electromagnetic fields—fields directly related to conditions under which ball lightning might form.

While Tesla failed to produce a definitive explanation for ball lightning, his experiments overwhelmingly contributed to our understanding of electrical phenomena. Furthermore, his insights influenced a host of classified governmental technological research and development.

Today, scientists still explore the mysteries of ball lightning using principles of plasma physics and electromagnetic theory, areas Tesla pioneered.

Earlier, I briefly mentioned what most people don't know. Tesla wirelessly transferred electrical energy into untethered lightbulbs he'd stuck in formation in the dirt outside of his Colorado lab. The basic principle behind this experiment was Tesla's work with what he called "earth currents", natural electric currents that flow through the ground. During thunderstorms, these currents become quite strong due to lightning strikes and atmospheric electrical activity. From Tesla's notes and accounts, he placed and spaced incandescent light bulbs partially buried, with just their glass portions exposed. The bulbs reportedly glowed without any direct wire connections, powered purely by ground currents induced during the storm.

Moving forward, ball lightning appears as a glowing, typically spherical object ranging from pea-size to several meters in diameter, though most reported sightings describe it as basketball-sized. Unlike regular lightning, it persists for several seconds to minutes and moves horizontally, often passing through solid objects like windows or walls without causing damage.

Documented accounts of ball lightning stretch back to ancient times but it's allegedly been difficult to capture and contain in a lab. However, in 2012, Chinese scientists accidentally harnessed ball lightning on spectrographic video during a thunderstorm study, providing the first *scientific* recording of the phenomenon.

Deep down in my neurosensing heart, I don't believe for a split atom that it took this long. Laboratory experiments by Brazilian researchers supported China's findings by creating similar phenomena using silicon wafers—and Dr. Matsumoto was producing micro ball lightning in his lab during the 1980's.

A couple of theories attempt to explain ball lightning:

1. **Vaporized Silicon Hypothesis:** suggests that when lightning strikes soil, it vaporizes silicon compounds which then oxidize and create a glowing ball of plasma.

2. **Microwave Cavity Theory**: proposes that electromagnetic energy becomes trapped in a spherical atmospheric cavity.

3. **Magnetic Knot Theory**: was developed by physicists at the University of Innsbruck. It proposes that ball lightning forms when normal lightning creates a magnetic "knot"—essentially a self-containing electromagnetic field that traps plasma. This might explain its stability and ability to move through obstacles.

4. **Oxidizing Theory:** Oxidizing air theory suggests that lightning breaks apart air molecules, creating a self-sustaining chemical

reaction that produces the glowing sphere. Some researchers have also proposed quantum mechanisms, suggesting that the phenomenon might involve quantum effects on a macroscopic scale.

What makes ball lightning so interesting is its reported behavior. Witnesses describe it moving against the wind, passing through solid materials, and occasionally causing damage when it dissipates. Some accounts mention a sulfurous smell associated with its presence.

One distinctive aspect of ball lightning is that it typically appears during thunderstorms but can also manifest indoors, sometimes entering buildings through windows or electrical outlets. Unlike normal lightning, it moves slowly and at eye level, making for a most memorable experience.

One of the earliest accounts of ball lightning happened in 1754 when George Wilhelm Richmann became the first person recorded to die while conducting lightning experiments. He was reportedly killed by what witnesses described as a blue ball of fire.

During World War II, numerous aircraft crews reported encounters with mysterious aerial phenomena known as "foo fighters," including B-29 crews who described glowing orbs that seemed to follow their planes. These puzzling sightings drew serious military attention as officials sought to understand the phenomenon. The term "foo fighters" was coined by Allied pilots to describe these strange lights or metallic-looking spheres that would trail or maneuver alongside their aircraft. The nickname is believed to have originated among the pilots of the 415th Night Fighter Squadron in late 1944.

Pilots reported seeing strange glowing objects that appeared to intelligently fly near or follow their aircraft, often matching speed and maneuvers. These objects were described as small, round, glowing red-white-orange objects. Usually seen at night during missions over Europe, particularly Germany, foo fighters were never intercepted.

Despite concerns that foo fighters might be advanced German weapons, no evidence ever appeared in the public domain to support this. The phenomena were blamed on optical illusions, ball lightning, ice crystals in clouds, pilot fatigue, illusions, or other weather phenomena. It's been said in some circles that the German scientists rounded up and sent to Russia after WWII, under *Operation Paper Clip,* were behind the metallic spheres, and efforts were made by the US to learn, build, and understand the wireless plasma-conscious technology—but again—I strive to master fiction.

In 1984, a particularly well-documented case occurred on a commercial airliner when ball lightning entered the cabin during flight, moved down the aisle, and eventually disappeared without causing harm. The multiple witness accounts from this incident provided valuable data for researchers.

Ball lightning typically exhibits these characteristics:
- Duration: Usually 1-10 seconds, though some cases last minutes
- Size: Most commonly 10-40 cm in diameter
- Color: Usually orange or yellow, but also reported in blue, green, and white
- Movement: Often moves horizontally, seemingly unaffected by wind
- Temperature: Witnesses report varying heat levels, from none to intense
- Sound: Sometimes accompanied by hissing or buzzing noises
- Disappearance: Can either fade away gradually or end in an explosion

Modern research has made significant progress in understanding ball lightning. The Joint Institute for High Temperatures in Russia created plasma balls that exhibit similar properties using microwave radiation.

In 2019, scientists at the University of Western Australia developed a theoretical framework explaining how ball lightning might maintain its shape through electromagnetic fields.

The development of high-speed spectrographic cameras has also been crucial in recent studies. The 2012 Chinese observation captured the spectrum of ball lightning, showing strong silicon lines, supporting the vaporized silicon theory.

While usually harmless, ball lightning has caused damage that has been documented:

- In 1936, ball lightning reportedly destroyed a church steeple in England
- A 1963 incident in Illinois involved ball lightning melting through a window screen without breaking the glass
- In 1994, ball lightning in Eastern Europe was documented burning through electrical equipment while leaving surrounding objects untouched

The most valuable recent contribution to research has been the increasing number of video recordings from smartphones and security cameras, providing reliable data that enhances historical eyewitness accounts.

Yet, studying ball lightning remains challenging because:

- It cannot be reliably predicted, nor can humans create it on demand in a natural setting
- Its rarity makes systematic observation difficult

- Laboratory attempts to recreate it may not represent the natural phenomenon
- The variety in witness accounts suggests multiple types might exist
- Measuring equipment must survive proximity to high-energy events

The phenomenon demonstrates how even in our technologically advanced age, nature still holds mysteries that challenge our understanding of fundamental physics.

Remember, plasma and ball lightning are forms of ionized gas. Plasma, by definition, is ionized gas where electrons are stripped from atoms, while ball lightning is considered a plasma phenomenon containing ionized particles.

Each emits light (or not) and electromagnetic radiation. Plasma naturally gives off light and other forms of electromagnetic energy as electrons recombine with ions, while ball lightning is observed as a glowing, luminous sphere possessed of similar electromagnetic processes. Each may be influenced by electromagnetic fields.

Notably, plasma *can be* contained and directed by magnetic fields, and ball lightning has been observed to move in ways suggesting electromagnetic interactions. Each involve high temperatures, but plasma exists at very high temperatures where atoms have enough energy to remain ionized—unless it's cold plasma. While the exact temperature of ball lightning is debated, the visible glow suggests significant heat energy.

Ultimately, neither plasma or ball lightning are well-understood as states of matter and remain somewhat mysterious. While many researchers believe ball lightning is a plasma phenomenon, its exact formation mechanism and composition are still subjects of ongoing research and debate. Since it's been postulated that different types of ball lightning might exist, the same can likely also be said for plasma.

CHAPTER ELEVEN

Quantum Quackery
& Other Scientific Fables
Zero-Point Energy,
Electromagnetic Fields,
and Anti-Gravities

S omeone once said that one of the greatest obstacles to discovery and innovation is not ignorance but the illusion of knowledge— and the things we *think* we know hinder what we *need* to learn. I wish I could remember who said this.

While it's fair to say that modern science has accomplished grand technological advances built on the minds, patents, and backs of great engineers and scientists, the scientific *establishment,* has largely ignored electricity as a significant factor of space.

Dr. Harold Aspden, whom I introduced to you in the last chapter, was an electrical engineer and physicist spent forty plus years spinning against the vortex of conventional physics. His *Energy Science Report No. Six* reads like a manifesto—a call to reconsider some of the most fundamental assumptions about quantum physics and how the universe works, including recalling Einstein's theory of relativity.

At the heart of Aspden's argument is a deceptively simple question regarding inertia: Why do objects resist being pushed or pulled? This force is something we experience every day, yet according to Aspden, physicists got it completely wrong.

The prevailing view when this paper was written (proposed by researchers Haisch, Rueda, and Puthoff) suggested that objects resist movement because they interact with an ocean of quantum energy, one that pervades empty space. Aspden proposed instead, that inertia is an intrinsic property of particles themselves—specifically related to their *electrical charge* and a desperate drive to conserve energy.

Aspden's theory is a radical reimagining of space. Rather than existing as an empty vacuum, he believed space is a dynamic system with three components: a structured lattice of electrical particles, a population of heavy "graviton" particles providing balance, and a sea of muons (heavy electrons) mediating between them. It's through the interplay of these components that he explains inertia, gravity, quantum mechanics, and even Einstein's famous $E=mc^2$ equation.

Aspden's frustration with the physics establishment is evident. He describes a "dam" built by mainstream physicists that blocks new ideas from flowing into "the knowledge stream." Despite having published comprehensive theories on gravity, inertia, and fundamental constants, he found himself consistently confronted with a lack of peer review and outright rejection. He was onto something.

Interestingly, everywhere I researched, I could find Dr. Hal Puthoff and company poo-pooing, debunking or discrediting alternative theories, energies, concepts, or inventions as huff and fluff or utter nonsense—and Puthoff has some high-level cred because he is super-smart and cloaked (I suspect) in a wardrobe of ultra, top secret cosmic wonder clearances.

Aspden (who died in London from complications of a stroke in 2011) suggested that the universe is far more dynamic than physicists imagine, with stable particles like protons and electrons constantly being created and destroyed. He argues that understanding this cosmic interplay between them could be key to tapping into new forms of energy.

Whether one agrees with Aspden's conclusions or not, his paper represents a fascinating attempt to rebuild physics from its established first principles. Like da Vinci, Bohm, and Tesla, Aspden challenges us to question even our most basic assumptions about how space, energy, and how the universe works.

Zero-point Energy

Zero-point energy (ZPE) is the lowest possible energy that a quantum mechanical system can have. It's the energy that remains even at absolute zero temperature (0 Kelvin) due to quantum fluctuations required by the Heisenberg uncertainty principle (where it's impossible to simultaneously know both the exact position and momentum of a quantum particle with arbitrary precision). This means that even in their ground state, quantum systems are never completely "still" but maintain a small amount of motion or energy.

The concept of "free energy" (in the sense of unlimited energy extraction from the vacuum) is not supported by mainstream physics. While zero-point energy is real, attempts to extract it as a power source are said to absolutely violate the fundamental laws of thermodynamics. This is different from the legitimate scientific concept of Gibbs free energy used in chemistry and thermodynamics. (Gibbs is a thermodynamic potential that measures the maximum amount of reversible work a system can perform at constant temperature and pressure, essentially indicating whether a chemical reaction or process will spontaneously occur.)

Again, we are cock-blocked with a straight and redacted arrow that instructs us to keep moving in a linear direction wearing blinders, limiting the efficiency of power stations, free-thinkers, car engines, and smart devices everywhere.

However, recent scientific breakthroughs in the public sector *are* harnessing ZPE through quantum vacuum fluctuation devices, to resonant cavity designs, which…drum roll please…enables propellant free propulsion! And don't even get me started on successful control of vacuum fluctuations using lasers. Yes, human beings are tapping ZPE fluctuations of the quantum vacuum.

But since I'm a fiction writer, let's compose a fairy-tale about how zero-point energy capture might work. Allow me to introduce Dr. Maxwell Entropy, who unfortunately either died in 1986 or simultaneously lives having transported himself to another dimension:

*In 1923, Dr. Maxwell "Max" Entropy was born to a quantum physicist mother and a stand-up comedian father. He was destined to find humor in the universe's most serious laws. After earning his PhD from MIT (Mostly Impossible Theories), he gained notoriety for his controversial paper "**Why The Second Law of Thermodynamics is More Like a Strong Suggestion.**"*

Despite his brilliant mind, Max was infamous for laboratory mishaps, including the time he accidentally teleported his granddaughter into a parallel universe. It was rumored that he once tried to power his car using only quantum vacuum fluctuations, resulting in a vehicle that simultaneously did and didn't move until someone observed it (his protegee, Stanley Meyer later built a car capable of running on any sort of water, whether normal, sea, or snow. Unfortunately, *authorities* labeled Meyer a fraud and the Belgian man who was with Meyer the night he died in the parking lot of a Cracker Barrel—who also happened to be heir to a vast vitamin drink fortune—swiftly left the US and years later submitted patents for…free energy technologies and other propulsion free systems.)

When not irritating the Nobel Committee with his interpretive dances about zero-point energy, Dr. Entropy could be found in his garden, attempting to explain quantum entanglement to Mr. Mew, his thoroughly unimpressed cat, (whose position and momentum were always entangled come dinner time.)

Entropy's classified ad for a lab assistant read: "Seeking non-Einstein type with low entropy and high energy. Must be comfortable with existing in multiple states simultaneously."

Dr. Entropy was also under the watchful eye of the CIA and every other three or four-letter agency that doesn't really exist. He discovered a way to create an asymmetric quantum vacuum fluctuation—essentially creating a "gradient" in the zero-point field that produces net directional force. His device precisely arranged arrays of nanoscale cavities that selectively amplified and suppressed different vacuum modes and nodes, creating this asymmetry.

Moreover, he challenged the second law of thermodynamics in quantum mechanics by demonstrating a system where quantum entanglement allows information to be preserved and extracted from what appears to be an increase in entropy. This showed, at the quantum level, certain reversible processes previously thought to increase entropy, maintaining quantum coherence in a way that preserves the system's information state.

"The second law assumes statistical independence of microscopic states," he said. "But I've found that quantum entanglement creates hidden correlations that preserve information we thought was lost to entropy. My experiments show that properly arranged quantum systems maintain

coherence and extract useful work from what appears to be disorder, as long as one accounts for quantum correlations."

As we make our way out of fiction and back to reality (however defined), it is important to remember that Zero-point energy is crucial in several areas of quantum physics:

- Understanding quantum vacuum fluctuations
- Explaining the Casimir effect (a quantum force between closely spaced conducting plates)
- Calculating ground state properties of quantum systems
- Studying quantum field theory and vacuum states

Zero-point energy also has several practical scientific and technological applications and implications:

- Semiconductor physics and device design
- Understanding chemical bonding and molecular stability
- Explaining superconductivity at low temperatures
- Developing quantum computing technologies
- Improving atomic clocks and precision measurements
- Studying van der Waals forces between molecules

The most significant practical impact of zero-point energy is in our understanding of how materials behave at the quantum level; that understanding has led to advances in electronics, materials science, and quantum technologies. However, we are reminded repeatedly—programmed really—that while zero-point energy is a fundamental aspect of quantum mechanics, it cannot be "harvested" as an energy source, despite claims to the contrary.

And why can't ZPE be harvested?

Zero-point energy can't be harvested, we're told, because it represents the absolute minimum energy state of a system. Energy cannot be extracted from a system that's already at its lowest possible energy state because it would violate the rock-solid laws of thermodynamics—like trying to retrieve that last dollop of toothpaste out of an empty tube. While the quantum vacuum is "bubbling" with virtual particles and fluctuations, this energy can't be extracted in a net-positive way. I remain skeptical.

As Sir Nicholas Winton once stated before saving over 600 children from the Holocaust before WWII, under *Operation Czech Kindertransport*, "If it's not impossible, there must be a way to do it."

Here's another speculative sci-fi concept for ZPE harvesting that, while not allegedly physically possible with our current understanding, again makes for interesting *fiction*:

"A Quantum Vacuum Cascade Collector"

We create an artificial quantum field gradient using overlapping containment fields (that manipulate quantum vacuum states) of stable exotic matter with negative energy density. This gradient would create a triangular "downhill slope" in spacetime, causing virtual particles from the quantum vacuum to temporarily manifest in a preferred direction (think of it like spacecraft rolling down the slope of a perfect right triangle).

These manifested particles would be caught in a cascade chamber lined with metamaterials, which would then capture and stabilize them before they 're-annihilate'. The stabilized particles would then be fed into a quantum engine that converts their mass directly into usable energy

Of course, we'd have to develop quantum metamaterials that can interact with virtual particles. Oh, wait... according to current research, certain types of quantum metamaterials, particularly those designed to exploit the "dynamical Casimir effect," can interact with virtual particles, converting them into real photons by harnessing the fluctuations of the quantum vacuum.

In 2018, a team of scientists of the Technion's Physics Department and Solid-State Institute and the Technion's Faculty of Mechanical Engineering developed a new field: quantum metamaterials.

Simply put, they used a dielectric *metasurface*, which acts differently in response to left- and right-handed polarized light (think of the Casimir effect using mirrors to reflect and exchange infinite energies between them, especially when compressed). This imposes polarized light on opposite phase fronts that look like vortices, one clockwise and one counterclockwise. The metasurface had to be nanofabricated from transparent materials or else it would have obliterated the quantum properties.

The team conducted two sets of experiments to generate entanglement between the spin and orbital angular momentum (OAM) of photons (light). Researchers shot a laser beam through a nonlinear crystal to create single photon pairs with zero OAM and each in linear polarization— a superposition of right-handed and left-handed circular polarization, which corresponds to positive and negative spin. What resulted was that the photon became entangled with the spin.

More recently, researchers from Northwestern University teleported a quantum state of light through existing fiberoptic cables handling internet traffic without interference. While this isn't physical teleportation, it is a matter of carefully selecting wavelengths to minimize scattering and to protect quantum integrity—and it is instantaneous, meaning it is faster than light speed.

Here's the secret: Harness neutrinos, plasma, and/or photons and understand vortexes to rule the world. These sorts of experiments pave the way for unbreakable quantum encryption and a host of other mostly top-secret applications—even if taxpayers everywhere foot the bill.

Rest assured, none of us everyday Jane and John worker bees will have a 'need-to-know' clearance until the technology is rendered obsolete and passed down to us—many moons from now—if ever.

But wait! There's more!

Scientists *have* transformed light into a super solid!

Way back in Chapter One, I mentioned that matter traditionally exists as solids, liquids, gases, or plasma. However, researchers have since uncovered even more exotic states of matter.

Scientists at Pennsylvania State University, for example, cooled helium to less than one-tenth of a degree above absolute zero (around -273°C) and believed they had stumbled upon a new state of matter: a "supersolid." Although initially uncertain about their findings, a portal had opened.

Shortly thereafter, teams from MIT in Cambridge, Massachusetts, and ETH Zurich in Switzerland, each using different methods, succeeded in creating supersolids through Bose-Einstein condensates.

Not to be outdone, a team led by scientists at Italy's National Research Council (CNR) achieved something even more astonishing: they created a supersolid using photons. While light had previously been turned into a superfluid, engineering a supersolid from light was unprecedented. Their groundbreaking experiment, detailed in *Nature*, involved using a patterned semiconductor and a laser to generate hybrid light-matter particles called polaritons, which then coalesced into a supersolid state. In simple terms, its light coupled with matter to create something entirely new.

This remarkable achievement not only expands our understanding of quantum phases of matter under non-equilibrium conditions but also paves the way for transformative advancements in quantum technologies.

Electromagnetic Fields

From holding atoms together to beaming power across continents (or galaxies), electromagnetic fields (EMF's) are as close to the supernatural as science gets. They've shaped our past, electrify our present, and promise a future straight out of a best-selling sci-fi novel.

Whether it's smartphone notifications or sunlight warming our face, EMFs are always there—an invisible symphony of frequency, energy, and vibration harmonizing sentient existence. So, the next time you scroll through social media or stare up at the stars looking for drones or UFOs, take a moment to appreciate the waves (and satellites) that make it all possible—and try to remember why you're here. One exists for more than just a job or paying the bills. And we are so much more than our egos and cravings. But I digress...

...Subsequently, EMFs exist everywhere in the universe from biological processes in living organisms to lightning, wi-fi networks, and atoms—and they've been here since the beginning of Earth time. EMFs appear to hold it *all* together, again making me question scientists' insistence that gravity rules. Likewise, electromagnetic fields are regions of space where electric and magnetic forces interact.

They consist of two interconnected components:
- **Electric fields**: Created by differences in electrical voltage/charge
- **Magnetic fields**: Created by moving electric charges (current)

These fields transport energy through space in the form of electromagnetic waves, which are oscillating electric and magnetic fields traveling together.

Three scientists majorly contributed to our understanding of EMFs:

- Michael Faraday (1791-1867): Discovered electromagnetic induction
- James Clerk Maxwell (1831-1879): Unified electricity and magnetism mathematically through equations and waves of oscillating electricity and predicted with some accuracy that magnetic fields travel in space at a particular speed, calculated close to the speed of light (sometimes his equations are jokingly considered a curse and deception for quantum physicists)
- Heinrich Hertz (1857-1894): Experimentally proved the existence of electromagnetic waves

EMFs are fundamental to nature and physics and its one of the four fundamental forces of nature (alongside weak, gravitational, and strong) because it holds atoms together and enables light propagation. It has also revolutionized modern life. EMF technology is responsible for power generation and distribution, telecommunications, electronics, wireless charging and navigation systems.

Furthermore, in biology, EMFs enable neural signaling, vision, and navigation (birds, bats, butterflies, and a host of other animals) and play a role in our healing (or harm).

Now, let's imagine space-based solar energy field systems are possible where, say, manned satellites orbiting Earth might collect solar energy in orbit using large arrays, perhaps converting solar energy to electricity using photovoltaic panels. Maybe these solar fields would beam energy back to Earth using microwave or laser transmission. They could receive and convert the energy at ground stations for distribution. If this was possible, perhaps these systems might operate in specific orbits, geostationary (35,786 km altitude) for constant positioning, medium Earth orbit for certain configurations, and special solar power satellite orbits designed for optimal coverage

The global privatized military-industrial-intelligence complex could plot out ground receiving stations (rectennae) located in remote areas with large land availability, regions with high energy demand, sites near existing power infrastructure or areas with minimal interference from weather.

The above-mentioned orbital solar fields were first proposed in the 1960's, with a few problematic prototypes in the 1980's. It appears, according to old USAF space technical reports found in a dusty Roswell archive, that the US lost interest for a while until we looked again in the 2000's. At the time, it was said that we'd have some better prototypes available to test by the mid-2030's. It also seems we were ahead of schedule, but again, nobody's talking—even if I slide proof in front of their eyes. It seems that ordinary vision isn't the problem as much as *first-eye* denial or my lack of need-to-know.

The awesome thing about collecting solar energy from space fields is the advantage of 24/7 large-scale power generation without the inconvenience of inclement weather or darkness. There would be zero emissions, minimal adversarial or ecosystem disruption, and a much-reduced need for energy storage or governmental-restricted eminent domain land acquisitions.

Subsequently, it would likely also provide black ops and covert research projects complete energy independence, remote power, and have a myriad of military applications. Of course, space debris would, I think, be a legitimate problem along with the extraordinary initial investment for infrastructure.

Perhaps some of the money for such a lofty project might have gone missing from the Pentagon, where it has failed six consecutive annual audits in a row, and is, officials say, years away from coming clean—I mean—producing a clean audit.

Since the IRS is stellar when it comes to conducting taxpayer audits on private citizens, maybe they should be called in to help the government find billions of dollars and years' worth of our missing tax money (This was written pre-DOGE). While black projects are privately said to covertly siphon tax dollars from purposely bloated agency budgets for clandestine projects it rarely, if ever, admits to doing so. Much of this money is transferred to the private sector under *'unacknowledged special access projects'*, which fall under proprietary or intellectual property not subject to Freedom of Information Act requests. Do not try this at home. Or if you do, don't tell a soul or try to patent it—unless you're part of the clandestine "in-crowd".

As almost everyone knows, more down-to-earth solar energy fields (also called solar farms or solar parks) are large-scale installations of solar panels designed to harvest sunlight and convert it to electricity. They are typically located in desert regions, and rural and semi-rural areas.

Alternatively, EMF applications in medical settings has shown therapeutic benefits:

- Improved bone healing through pulsed electromagnetic field therapy (PEMF) in fracture treatment
- Reduced inflammation and pain management when used in physical therapy settings
- Potential mood improvements from transcranial magnetic stimulation (TMS) for treating depression
- Some evidence for improved circulation and tissue regeneration in controlled medical applications

Unfortunately, EMFs can also have negative effects too:

- Sleep disruption from exposure to EMFs, particularly from electronic devices at night
- Potential stress responses in some sensitive individuals, including headaches and fatigue
- Concerns about prolonged exposure to high-frequency EMFs from various electronic devices

Now, no book with a chapter on electromagnetic frequencies would be complete without mentioning American inventor Royal Raymond Rife (1888-1971). He claimed in the 1920s and 1930s that he could destroy cancer cells and pathogens using specific electromagnetic frequencies, which he called a "Beam Ray Device" or "Rife Machine." He theorized that every organism had its own unique resonant frequency, and that using the correct frequency could destroy harmful organisms while leaving healthy tissue unharmed (more about resonant frequencies in Chapter Twelve).

Rife initially gained support from the medical community. He performed demonstrations at his lab in San Diego and worked with some respected doctors, including Dr. Milbank Johnson of USC, who ran clinical trials of the device.

Unfortunately, like Tesla and Bohm, Rife experienced major rejection and was discredited. The medical establishment turned its back on Rife and one of his investors Philip Hoyland, sued Beam Ray Corporation (Rife's company), leading to its collapse (the exact same thing happened years later to Ohioan Stanley Meyer and his water-powered car).

Rife's original device was reportedly destroyed in a fire, though there are conflicting accounts about this. His laboratory notes and many of his microscopes were also lost or destroyed under disputed circumstances.

Not surprisingly, Rife developed a serious alcohol problem and isolated himself from the scientific community. His reputation suffered further damage after the Journal of the American Medical Association published an article labeling his work a hoax. Rife died in 1971 in El Cajon, California, largely forgotten. After his death, others tried to market variations of Rife's technology, but the FDA swiftly stepped in to act against them for making 'unproven medical claims'.

From a more spiritual perspective, a few shamans, yogis, religious leaders, and priestesses believe natural Earth EMFs enhance meditation and grounding while others express concerns about artificial EMFs disrupting human energy fields. Many traditional practices emphasize connecting with natural electromagnetic fields of the Earth for wellness.

For example, grounding/earthing exercises in natural electromagnetic fields, working with Earth's Schumann resonance (7.83 Hz) through meditation, movement of subtle or natural energy fields (tai-chi or qi gong), holotropic breathing exercises, connecting with natural magnetic nodes or power spots, sunrise/sunset practices when Earth's EMF field undergoes natural shifts, or using specific electromagnetic frequencies believed to correspond to different states of consciousness.

Sound healing is another practice combined with EMF awareness, as is biorhythm studies or leveraging the magical properties of crystals and copper.

When it comes to EMF spiritual work, distance matters because EMF strength decreases significantly with increased distance from the source. Additionally, duration of exposure impacts potential effects. Also, artificial sources might have different impacts, and it's noted that individual sensitivity varies.

There is scientific evidence showing potential harmful effects from artificial EMFs such as DNA damage and oxidative stress from high-frequency EMFs in laboratory studies. Other research shows cellular stress responses, particularly with prolonged exposure to strong fields, sleep disruption from nighttime exposure to artificial EMFs, especially from blue light and electronic devices, and heat-related tissue damage from very high-intensity EMF exposure (the basis for microwave safety standards).

There might even be evidence to suggest negative impacts on fertility and reproductive health, particularly from carrying phones in closely kept purses or pockets, neurological effects including headaches and concentration difficulties, and disruption of circadian rhythms affecting hormone production. Could it be that falling birthrates in first world countries might be linked, in small or large part, to our dependence upon and exposure to technology?

During EMF studies that have (regrettably) included animals, there exists concerning evidence for carcinogenic effects (WHO's International Agency for Research on Cancer classifies RF EMFs as "possibly carcinogenic"). There are also impacts on childhood development from prenatal exposure, and effects on wildlife, particularly on bird and sea turtle navigation and insect behavior.

The scientific consensus is that while risk exists (with everything), high-intensity exposure is probably the worst.

Fortunately, most everyday EMF exposures, we're told, fall within generally accepted safety range limits. However, research continues to examine long-term effects and potential mechanisms of EMF levels. Again, the debate continues.

CHAPTER TWELVE

Casimir is CERN-tainly Not What I Expected

Quantum Fluctuations and How I Learned to Love the Vacuum

In 1948, Hendrik Casimir predicted that if you place two uncharged metal plates extremely close together in a vacuum, they will attract each other. This was experimentally proven in 1997 by Steve Lamoreaux. It works because in normal space, virtual particles of all wavelengths are constantly popping in and out of existence.

When you place the plates very close together (less than 1 micron apart) only certain wavelengths exist between the plates (like a guitar string that can only vibrate at specific frequencies) but outside of the plates, all wavelengths can exist. This creates a pressure difference, pushing the plates together. The force is tiny but measurable, about 1/30,000th of the weight of an ant.

Additionally, atoms in excited states will spontaneously emit light even in complete isolation. This happens because the zero-point field interacts with the atom's electrons and without the zero-point field, excited atoms would remain excited indefinitely.

Moreover, Van der Waals Forces are the weak forces that let geckos stick to walls and part of these forces comes from interaction with the zero-point field. The quantum vacuum helps molecules stick together.

Alternatively, while the regular Casimir Effect uses static plates, the Dynamic Casimir Effect experiment involves rapidly moving mirrors. In 2011, scientists at Chalmers University created an artificial version using a superconducting circuit. A superconducting circuit contains a boundary condition (like a mirror) that can be moved at extremely high speeds and movement at high enough speed (about 5% of light speed), causes the vacuum to release real photons.

This is bit like shaking a feather duster where the motion causes dust (in this case, virtual particles) to become *real*—and it proves that the vacuum (which might be fields of waves rather than grains) can be "excited" into producing actual particles.

Interestingly, and building upon Casimir, Willis Lamb discovered what is known as the Lamb Shift in 1947. He discovered that two energy levels in hydrogen atoms that should be identical have a tiny difference—and this difference is caused by the electron interacting with virtual particles in the quantum vacuum. This experiment was one of the first proofs that the vacuum isn't truly empty.

Moreover, and years later, scientists created "squeezed vacuum states" where quantum fluctuations are reduced below normal vacuum levels in some directions. This has applications in improving the sensitivity of gravitational wave detectors. Researchers continue to examine the effects of this research for exploring further and faster applications in quantum computing and quantum cryptography.

Hawking Radiation also shows us that the vacuum of space is not truly empty because black holes are predicted to emit radiation due to quantum effects at their event horizon—virtual particle pairs are created, with one falling in and one escaping, causing black holes to slowly evaporate. While not directly observed, this process relies on the same quantum vacuum effects we see in labs.

What's CERN got to do with it?

CERN, French Conseil Européen pour la Recherche Nucléaire, (European Organization for Nuclear Research) operates the world's largest particle physics laboratory near Geneva, Switzerland.

The "God Particle" is the popular yet scientifically controversial nickname for the Higgs boson, discovered at CERN in 2012. The Higgs boson is fundamental to modern physics because it gives other particles their mass through the *Higgs* field. The Higgs field is like other fields in physics (such as electromagnetic fields), but with a crucial difference: it has a **non-zero value** even in empty space. This non-zero value is called its "vacuum expectation value." I envision it as a superconductor.

This property is what allows it to give mass to particles even when they're just moving through seemingly empty space. Its discovery confirmed a major prediction of the Standard Model of particle physics, solving a decades-old mystery about how particles acquire mass. The Higgs boson is essentially a ripple or excitation in the Higgs field—like how a wave is a disturbance in water, except that we can't *see* the waves at the subatomic level the way we do in a body of water.

CERN found the Higgs boson using the Large Hadron Collider (LHC), a 27-kilometer circular tunnel that accelerates protons to nearly light speed and collides them. The collisions create conditions like those just after the Big Bang, allowing scientists to observe fundamental particles and forces.

The nickname "God Particle" came from physicist Leon Lederman, though many scientists dislike the term. The actual Higgs boson is a subatomic particle that exists for only a fraction of a second before decaying into other particles.

But what I find even more fascinating is that as particles in the accelerator approach the speed of light, they *appear* to become shorter in the direction of motion. From the proton's frame of reference, something interesting happens: the entire world around it, including the LHC tunnel itself, contracts—not in a physical way but quantumly. So, from the proton's point of view, it's traveling around a suddenly drastically reduced circumference of approximately 3.62 meters instead of the 27 kilometers! This is related to what's known as the *Lorentz* contraction.

At the speeds achieved in the LHC (99.999999% the speed of light), the length contraction effect is significant. The protons themselves would appear to be compressed by a factor of about 7,460 from our perspective. However, this doesn't affect the actual physical structure or dimensions of the accelerator tunnel.

This quantum-level contraction is one of the most mind-bending aspects of special relativity, that the same *physical* situation can appear very different depending on our frame of reference, yet both *perspectives* are equally valid.

Breaking the Quantum Code: Neutrinos Revisited

As previously discussed, neutrinos are some of the most abundant yet elusive particles in the universe. They're elementary particles, meaning they're not made up of smaller components but are considered fundamental building blocks of matter.

Remember, neutrinos have several unique characteristics:

- Extremely negligible mass
- No electric charge

- Travel at nearly the speed of light
- Very weak interactions with other matter (trillions pass through your body every second)

And neutrinos are created through nuclear fusion in stars (including our Sun), radioactive decay, nuclear reactions (both in power plants and naturally occurring), cosmic ray interactions with the atmosphere and Supernova explosions.

Neutrino Relationships with the Zero-Point Field

The connection between neutrinos and the zero-point field is still a matter of public theoretical research. Neutrinos participate in quantum vacuum fluctuations, the constant appearance and disappearance of virtual particle-antiparticle pairs in empty space—they pop in and out of existence. Even though their contribution is smaller than other particles due to their weak interactions, they're part of the *quantum foam,* which characterizes the *vacuum* state of space.

Neutrinos do interact with the Higgs field, but this field does not directly give them their tiny mass, suggesting another medium may be responsible. Remember, neutrinos are primarily "left-handed" particles and are not able to switch handedness under the Standard Model of physics.

This weak coupling to the Higgs field is one of the great puzzles of particle physics. A hint of a right-handed *sterile* neutrino may exist, lurking somewhere—but in the realm of publicly available science it has not yet been discovered.

Yet due to their weak interactions, neutrinos can maintain quantum coherence over longer distances than many other particles. This has interesting implications for quantum mechanics and has led to phenomena like neutrino oscillations, where neutrinos can allegedly change form between their three "flavors" (electron, muon, and tau neutrinos).

One fascinating aspect of neutrinos is their potential role in several unsolved mysteries of physics:

- The matter-antimatter asymmetry in the universe
- Dark matter (though they're not the primary component)
- The exact mechanism of supernova explosions

Furthermore, the phenomenon of neutrino oscillation (where neutrinos change between their three tau, muon, electron flavors, and possibly sterile neutrinos) demonstrates a unique quantum mechanical behavior that affects the energy distribution in quantum fields (I doubt neutrinos change flavors as much as existing between dimensions in superposition).

Yet the most fascinating aspect is how neutrinos, despite their weak interactions, still play a crucial role in maintaining the quantum energy balance of the universe. They act as subtle but highly important contributors to the overall quantum energy landscape—like relatively obscure Sci-Fi writers.

Substantially, neutrinos can maintain quantum coherence over long distances and are capable of near light-speed travel. They have an ability to penetrate, according to current scientific understanding, all matter. Some speculate that neutrinos might play a role in non-local information and energy transfer in consciousness and that their quantum properties might contribute to a universal field that some theorize could be related to consciousness.

As we explore theoretical mechanisms such as Quantum Coherence (maintained quantum states in biological structures), Non-local Connections (quantum entanglement potentially enabling instant information transfer), Field Interactions (consciousness possibly emerging from interactions with quantum fields), Information Processing (quantum effects potentially enabling complex information processing in the brain)—we arrive at a horizon where neutrinos might serve as potential carriers of consciousness-related information and somehow employ the zero-point field as a universal memory bank.

CHAPTER THIRTEEN

The Devil's Frequency
When Harmonics Go HAARP-Wire, Call in the Resonance

The ubiquity of *resonance* suggests that it is one of the universe's fundamental organizing principles, helping create order and pattern from chaos. Resonance occurs when the frequency of an initial object's vibration matches the resonant frequency or natural frequency of the second object.

Two people can also experience a kind of resonance when their thoughts, emotions, or intentions align, creating a deep sense of connection or mutual understanding. This often occurs through subtle, nonverbal exchanges—body language, fleeting glances, or micro-expressions that transmit meaning beyond words. It's an energetic communication, one *felt* more than analyzed. Yet, because this type of resonance—rooted in subjective experience—can't be easily quantified or reduced to discrete units, it is often dismissed or overlooked by conventional science.

It's intriguing how resonance connects phenomena from quantum mechanics to cosmic structures, and from lifeless matter to biological and conscious experience. Even empty space itself appears to have resonant modes within both quantum field fluctuations and vacuum energy states. Its pervasiveness points to deeper quantum-like patterns in human understanding about the nature of reality and makes those microtubules within our brain much more interesting for serious study.

Here's what's truly riveting: Researchers at the Institute for Materials Research at Tohoku University, in collaboration with the Japan Atomic Energy Agency and the RIKEN Center for Emergent Matter Science, made a groundbreaking discovery. While experimenting with nanoscale magnetic materials, they observed an *asymmetric diffraction pattern* in surface acoustic waves—a phenomenon previously seen only in optics.

This unexpected behavior suggests that sound waves, like light, can be manipulated at the quantum level. The implications are enormous: it could revolutionize both classical and quantum communication technologies and possibly offer a new method for transmitting information across dimensions—hinting at a future where even telepathic-like communication may be scientifically viable.

Consequently, ancient civilizations discovered and used acoustic resonance in musical instruments and architecture, like Greek amphitheaters. Chinese craftsmen used bronze bells as early as 2000 BCE to communicate with ancestral spirits and create a sacred soundscape. Pythagoras discovered mathematical relationships in musical harmonics and resonant strings, and Islamic scholars like Al-Farabi advanced the mathematical understanding of musical resonance.

In fact, our favorite curiosity seeker, da Vinci also observed and documented sympathetic resonance between musical strings, noting that when one string is plucked, other strings tuned to harmonically related frequencies began to vibrate on their own. This is one of the earliest documented observations of resonant coupling.

The Labyrinth Group (not its real name)—an alleged offshoot of the Advanced Contact Intelligence Organization under the National Security Agency—took the concept of resonance to a whole new level when its project leader, A.R. Bordon (from either Missouri or Brazil) (aka "Fifteen"), and his team of researchers converged on the Ancient Arrow Site, roughly eighty-three miles northeast of Chaco Canyon, New Mexico. This all began around 1972, after a group of young hikers *ALLEGEDLY* stumbled upon something... otherworldly.

The caves at the Ancient Arrow Site were said to contain twenty-three chambers filled with strange thousands of years-old artifacts—paintings, poetry, and, most intriguingly, an 8,000-page coded disc written in a language that initially at least, no one could decipher. Six to twelve other sites around the globe are also said to exist and have not yet been found (unless we count the recent Sidar discovery by Italian researchers under the Pyramids at Giza—more research is warranted).

Spiraling further into this rabbit hole of half-truths, suffice it to say: the chambers were stamped with an "above top-secret" classification. Why? Because when the artifacts were used together their resonances opened vortexes to other dimensions—either to encourage alien contact or to directly experience communication with interdimensional beings. Taken together as a whole, the 'music', art, and objects inside the caverns were claimed to protect humanity or elevate consciousness beyond its primal state. The NSA isn't talking and will neither confirm nor deny the existence of these caves. Other researchers vehemently claim the entire story is a laughable hoax.

What happened to the linguist tasked with deciphering the disc—Dr. J. Neruda—is anyone's guess. He claimed in an interview with a journalist originally named Ana, that he succeeded in translating a portion of the disc. He supposedly defected from the Labyrinth Group a few years into the project and vanished after he outed the Labyrinth Group's work. He also expressed displeasure over how one remote viewer, Samantha, was blank slated, her mind wiped clean because A.R. Bordon was threatened or resentful about her high-level remote viewing capabilities.

As for Bordon, he reportedly passed away "a while ago" around the age of ninety. A LinkedIn profile still exists for him as of this writing, as does a Facebook Group dedicated to his work and his "passing on in spirit".

Remember, the more ridiculous a story sounds, the easier it is to bury.

What I do know based on research and after reading six of A.R. Bordon's nearly out-of-print books and numerous papers, is that clandestine agencies have been studying off-limit archeological sites, mind control, music, art, quantum physics, plasma, free energy, interdimensionals, electromagnetics, and time travel—and the original story is quite different and much more believable than the current version.

Years later, a group calling themselves the WingMakers—previously untraceable through any IP address—surfaced, claiming the entire Chaco Canyon site and its discovery, and everything about the Labyrinth Group or the AICO was a fictional narrative device. Funny, they failed to mention COIST, which was likely the true name of the group—except I find no record or mention of COIST anywhere—except in an old out-of-print book written years ago by...A.R. Bordon.

The Labyrinth Group, once based in Virgina, not California, is now claimed to be a provocation to expand human thought brought to us by a talented artist named James Mahu (In Hawaiian, Mahu means having both male and female spirit. Mahu were teachers, healers, and keepers of sacred knowledge).

The Chaco Canyon Artifacts/Vertical Time Travel/Psychotronic Generator story of the Labyrinth Group within the AICO was edited and re-edited so many times, it's become a mystery unto itself. The original WingMakers site, which picks up and alters the tale, was built by Mark Hempel after he received a package of detailed instructions on exactly how to proceed.

A more believable and publicly available ancient site you can visit is Rosslyn Chapel, founded in 1446 near Edinburgh, Scotland. It became famous for its intricate carvings and association with Dan Brown's novel, *The Da Vinci Code*.

Researchers Thomas and Stuart Mitchell (father and son) spent years deciphering a musical score encoded within the Rosslyn chapel's stonework. They theorized that carvings on the arches and stone cubes contain symbols related to Chladni patterns (visual representations of vibrational modes on a flat surface, which depict musical frequencies). Years later, a beautiful song based on their work was sung by a choir within the chapel. This leads me to question how many other sonic geometry codes freemasons openly hid within divine or sacred architecture?

For the sake of ease, we can break down resonances into the following categories:

Cosmic

- Orbital resonances between planets, moons, and asteroids
- Gravitational waves rippling through spacetime
- Resonant frequencies in stellar oscillations
- Galaxy rotation and spiral arm patterns

Atomic/Quantum

- Electron orbital resonances
- Nuclear magnetic resonance
- Quantum wave functions
- Atomic energy level transitions

Biological

- DNA and protein vibrations
- Cell membrane oscillations
- Neural firing patterns
- Heart rhythms and circadian cycles

Physical/Chemical

- Chemical bonds and molecular vibrations
- Crystal lattice vibrations (phonons)
- Light-matter interactions
- Sound wave propagation

Mechanical Resonance

- Mechanical resonance occurs when an external force matches a system's natural frequency. This includes structural resonances in buildings and bridges, and acoustic resonance in musical instruments. Mechanical resonance always makes me think of a soprano opera singer using her voice to shatter a crystal glass, or ancient engineers levitating large boulders to build pyramids.

Electromagnetic Resonance (ER)

- This includes electrical circuits, where inductors and capacitors exchange energy at specific frequencies. ER is crucial for radio transmission, MRI machines, and wireless charging. Cavity resonators and waveguides also use electromagnetic resonance.

Quantum Mechanical Resonance

- Nuclear magnetic resonance (NMR) used in medical imaging
- Electron spin resonance in materials science
- Feshbach resonances in ultracold atomic physics, which are quite useful for studying quantum phenomena like superfluidity. It allows for precise control between atoms (another accidental exposure, in which I'll leave the reader a mystery to solve).

Chemical Resonance

- Also called mesomerism, this describes the delocalization of electrons in molecules, particularly important in organic chemistry and biochemistry. Benzene's structure is a classic example.

  Consider an interesting thought experiment: If we could control this resonant process within a subconscious altered state, we'd be able to make matter exist at will in multiple locations simultaneously (at a quantum level), or "collapse" this superposition at a desired location, effectively moving the matter—except Heisenberg's uncertainty principle means we'd never precisely know both position and momentum and the energy required would be enormous. Ah, well—perhaps it could be used to temporarily "loosen" the quantum structure of matter to facilitate transfer through some kind of quantum tunnel or wormhole—but it's a sci-fi stretch.

Optical Resonance

- Occurs in laser cavities and optical fibers
- Surface plasmon resonance used in biosensors

- Photonic crystals and metamaterials, which both control the propagation of electromagnetic waves. Metamaterials can be used to cloak objects while photonic crystals may be used for crystal lasers or optical computers

Nuclear Resonance

Nuclear resonance happens when an atomic nucleus absorbs energy at exactly the right frequency to transition between specific quantum energy states, like how a tuning fork resonates with sound waves of its natural frequency. This can lead to something called the Mössbauer Effect: This specific type of nuclear resonance, where atomic nuclei in a solid absorbs and emits gamma radiation with no recoil loss of energy, enables ***extremely precise measurements***. *Doh!* We have another accidental exposure (she says with a smile).

Planetary Resonances

Planetary resonances are orbital relationships between celestial bodies where their orbital periods form simple numerical ratios. Orbital resonance occurs when two objects' orbital periods are related by a ratio of small integers. For example, a 2:1 resonance means one object completes two orbits at the same time another completes one orbit.

In our solar system, Neptune and Pluto have a 3:2 resonance, meaning that for every three orbits Neptune makes around the Sun, Pluto makes two orbits. This helps stabilize Pluto's orbit despite its path crossing Neptune's.

Jupiter's largest moons demonstrate multiple resonances. Io, Europa, and Ganymede are locked in a 4:2:1 resonance, meaning Io completes four orbits for every two of Europa's and one of Ganymede's.

Resonances stabilize orbits, helping maintain optimal orbital patterns over long periods—but they can also commit orbital migration, where gravitational interactions (decoherence) gradually change the orbits of bodies over time, which might lead to chaotic behavior or even a planet's ejection from a solar system.

The study of resonances is crucial in understanding the formation and evolution of our solar system, particularly in explaining the current orbital architecture of planets and their satellites.

Aligning with natural frequencies, rather than fighting them, tends to create the *best* conditions for resonance. Resonance allows small, well-timed inputs to create large effects by matching the natural rhythm or frequency of a system. It's as if the system "prefers" to respond at certain frequencies, and when those frequencies are matched, we get enhanced effects.

Resonance turns out to be incredibly powerful and appears throughout nature, from the smallest quantum scales to the largest astronomical systems. It also lends credence to Tesla's wise words that "*...if we hope to understand the secrets of the universe we must think in terms of energy, frequency and vibration.*"

Biological and Consciousness Connections to Resonance

Neural oscillations (brain waves) operate through resonant frequencies, with different states of consciousness associated with specific frequency bands (alpha, beta, theta, delta).

The 7.83 Hz Schumann resonance of Earth's electromagnetic field closely matches human alpha brain waves, which some researchers hypothesize influence consciousness and circadian rhythms.

Additionally, the cochlea in our ears uses mechanical resonance to break down complex sounds into frequency components, enabling our rich experience of music and speech.

Spiritual and Meditative Resonance Practices

Chanting and mantras often use specific frequencies thought to *resonate* with different aspects of consciousness or energy centers. For example, Tibetan singing bowls create complex resonant frequencies used in meditation and healing practices. Sacred spaces like cathedrals are often designed with specific acoustic resonances to create uplifting psychological, spiritual, and emotional effects and many ancient traditions use rhythmic drumming to induce altered states.

Some Scientific/Technological Applications of Resonance

- Medical imaging (MRI) uses magnetic resonance of hydrogen atoms.
- Quantum computers utilize quantum resonances for computation.
- Wireless power transfer is achieved through resonant coupling.
- Resonant circuits are used in all modern electronics and communications.
- Lasers operate through optical resonance.
- Gravitational wave detectors use incredibly precise resonant systems.
- Energy harvesting is achieved through mechanical resonance.

The emergence of resonant phenomena across biology, consciousness, and technology again suggests a fundamental organizing principle in nature. Resonances provide a bridge between physical vibrations and subjective experience—for example, how specific sound frequencies reliably induce heightened or lowered emotional states or levels of consciousness.

At this moment, I am listening to Chopin on a soundtrack entitled *Music for Cats*. While I can't yet tell if Schroedinger's cat or my writing resonates with the universe or tethers my attempt at some grand unification of science with the sacred, Mr. Mew, my typically nosy, laptop stomping house panther, seems to *resonate* with the music. Today, at least, he is lulled into a state of abnormally good behavior.

Earth's Frequencies

Winfried Otto Schumann (1888-1974) was a German physicist who, in 1952, predicted and mathematically described the global electromagnetic resonance phenomenon that now bears his name. He forecast the existence of resonant electromagnetic waves in Earth's atmosphere while teaching at the Technical University of Munich.

Schumann's discovery was largely theoretical at first and his actual confirmation of these resonances wasn't achieved until the early 1960s, almost ten years later, primarily through the experiments of his student Herbert König.

The Schumann resonance occurs because the space between Earth's surface and the ionosphere (about 60 km up) acts like a giant spherical cavity. Lightning strikes (which happen about 50 times per second globally) excite this cavity, creating standing electromagnetic waves. The fundamental frequency is approximately 7.83 Hz (close to human alpha brain waves, which are 8-13Hz), with harmonics at:
- 14.3 Hz
- 20.8 Hz

- 27.3 Hz
- 33.8 Hz

So, imagine the Earth and its ionosphere as a colossal spherical drum—or better yet, a planetary-scale Tesla coil. Every lightning strike is like a mallet to that drum, sending ripples of electromagnetic energy skipping around the globe. These impulses reverberate between the Earth's surface and the ionosphere, forming resonant standing wave patterns at precise frequencies—phenomena we now recognize as *Schumann resonances*.

To think these planetary vibrations don't affect the human mind and body—or the rhythms of animals and plants—seems implausible. We are, after all, electromagnetic beings living inside a resonant chamber. It's also interesting to note that the geomagnetic field has various frequencies associated with interactions between the Earth's magnetosphere and solar wind. And some research suggests that these frequencies influence cellular calcium ion movement and other biochemical processes that are fundamental to life.

Similarly, other research suggests correlations between significant geomagnetic disturbances that include increased rates of cardiovascular issues, changes in blood pressure and altered melatonin production.

Frighteningly, the weakening of Earth's magnetic field over time will lead to increased radiation exposure from space, frequent disruptions to beneficial frequencies, and greater vulnerability to solar storms. Let's hope this happens after humans and other animals have long ceased to exist—but I fear its already begun—and we're not being told.

When we reflect on our biological evolution—and my own hope for humanity's future, which admittedly resides more in the realm of compassion than locked inside a box of strict science—it seems clear that the stability of Earth's magnetic field has been essential. It has likely played a vital role in nurturing life, allowing organisms to evolve and adapt to the planet's resonant frequencies over millions of years.

Energy Harmonics

Energy harmonics are like invisible ripples in our electrical systems—waves that ride on top of the normal power frequency (usually 50 or 60 Hz), but at higher, repeating intervals. These aren't just technical quirks; they have real effects on both machines *and* living beings.

Some harmonics are disruptive, contributing to fatigue, insomnia, or even health issues for us and our pets. Others, however, may produce benefits—like how certain electromagnetic frequencies are used in therapies to promote healing or improve mood. Comparable to tuning an instrument, the energy flowing through our environment affects us more than we realize.

For our purposes think of electrical current as a smooth wave moving through power lines. The main wave, called the fundamental frequency, is like the bass beat of a song. For example, the 2nd harmonic is twice the fundamental frequency and so on…These harmonics can cause problems in electrical systems, like overheating, power failures, or making power less efficient. Engineers work to minimize energy harmonics to keep power grids stable and devices running smoothly.

Human Adaptations of Electrical Harmonic Principles

Let's consider playfully adapting the energy harmonics template used by electrical engineers for power grids as an overlay upon the human body's energy systems.

Installing *harmonic filters* (a wearable device, frequency mapping, meditation etc.) might involve developing methods to screen out disruptive environmental frequencies, create "energetic boundaries" or protective personal spaces and shields, and translate to techniques for filtering out unwanted mental/emotional/energy patterns.

"Phase-shifting transformers" could help a human shift between different states of consciousness or develop methods for adapting to different energy levels throughout the day. Potential applications for helping balance various aspects of human energy systems might be found in "Active harmonic conditioners", which monitor and adjust personal energy states to actively maintain homeostasis states.

Human system ratings could be employed to understand individual capacity and tolerance levels, determine optimal frequencies/resonance/harmonics for different people or offer personalized approaches rather than *one-size-fits-all* wellness.

Then, we might use "Power quality analyzers" to translate methods for measuring subtle energy states, monitor physiological responses, and maintain optimal energy levels.

What About Scalar Waves?

Scalar waves, often associated with Nikola Tesla, are a sort of inter or multi-dimensional electromagnetic wave. Scalar waves differ from radio waves or light and are described as longitudinal waves of potential that have magnitude but not defined direction.

Scalar waves are believed to exist as stationary energy patterns that oscillate longitudinally along their propagation direction. Some theories suggest Scalar waves produce anomalous effects, particularly during solar eclipses, can influence Faraday-enclosed detectors, induce sidereal wave variations in torque pendulums, and generate unexpected voltages in dielectrics.

Coincidentally, one patent by Hal Putoff, et al, 2016 (Communication Method and Apparatus with Signals Comprising **Scalar** and Vector Potentials without Electromagnetic **Fields**," **U.S.** Pat. No. 5,845,220) claims, in short, that scalar fields can 'break away' from their antennae structure and propagate as independent entities! Relatedly, Scalar waves are reputed to be impervious to electromagnetic shielding.

While studying the rhythmic patterns of lightning strikes in Colorado, Nikola Tesla observed a remarkable phenomenon: the Earth itself responded with resonant electrical echoes. These impulses traveled not just across the surface but *longitudinally*—through the Earth— to the antipode, where they reflected and returned. This suggested to Tesla that the Earth and its upper atmosphere could together act as a colossal spherical capacitor, capable of storing and discharging high-voltage electric currents.

Even more intriguing was Tesla's realization that under certain conditions—particularly at high frequencies and voltages—air behaves paradoxically: it can act as both an excellent insulator and a superb conductor. This dual behavior more than hints at mechanisms that transcend conventional electromagnetic theory—and drags most scientists kicking and screaming, straight into quantum realms.

Some researchers now speculate that Tesla may have unknowingly been tapping into scalar wave phenomena—non-Hertzian waves theorized to propagate through the vacuum or *aether* without needing a conventional medium.

Additionally, such high current voltages likely generated radiation objects like 'plasmoids'. Notably, physicists and academics do not understand or accept Tesla's concept of longitudinal waves—even though such waves have long been proposed as a long-distance communication method resistant to shielding—even underwater.

The Ionosphere

The ionosphere is a layer of Earth's upper atmosphere, extending from about 60-1000 km above the surface. It's characterized by ionized gases (atoms and molecules that have lost or gained electrons due to solar radiation). This ionization creates multiple distinct layers (D, E, and F layers) that vary in density and height throughout the day and seasons.

The ionosphere is important to life and human experience because it reflects and refracts radio waves, enabling long-distance radio communication. It protects Earth from harmful solar radiation. It plays a role in creating aurora borealis/australis, helps maintain Earth's electromagnetic field and is critical for satellite communications and GPS.

Moreover, you can receive more stations at night with a shortwave radio because the ionosphere reflects radio waves more effectively. This allows signals from distant stations to travel further by "bouncing" off the ionosphere and reaching your location, a phenomenon called "skywave propagation".

HAARP (High-frequency Active Auroral Research Program)

HAARP is a scientific research facility in Alaska that studies the ionosphere. Its main component is the Ionospheric Research Instrument (IRI), a large field array of high-power radio transmitters that beam focused energy into the ionosphere. It's been said that its applications include:

- The study of ionospheric physics
- Research into radio propagation
- Analyzing aurora formation
- Testing communication improvements
- Investigating natural ionospheric processes

Conspiracy theories abound regarding HAARP (High-Frequency Active Auroral Research Program). It is claimed that this research facility is used to "push out" the ionosphere to enhance the Earth's protective shield, but instead, might have accidentally or intentionally manipulated weather patterns, causing natural disasters like earthquakes, mud slides, and hurricanes. What is true is that HAARP scientists have sent high frequencies into the ionosphere.

The ionosphere protects us by working with Earth's magnetic field and other atmospheric layers to maintain conditions suitable for life. In 1993, the US sent billions of watts into the ionosphere so it would vibrate back to Earth and our government could explore tunnels and search for deeply buried oil or natural gas.

When humans attempt to push out the ionosphere through irradiation, what happens? Weather is allegedly seriously disrupted because what goes up, we're told, must come down. Did billions of watts ricochet back to Earth and we're paying for it today with crazy, unpredictable weather patterns? People aren't talking or the answer is a reverberating *"NO!"*.

Ironically, ionospheric warfare studies were also conducted in the former Soviet Union and the US. The public story claims these studies were done to measure heat on satellites, communication disruption, and wave propagation effects among others. The results of the studies measuring irradiation levels of disease, stress, passivity and 'hysteria', *because* of these experiments, **remain classified**.

Other wild theories claim that HAARP is secretly used to control the minds of the masses or to purposely interfere with communication systems. When I'm in Washington DC, and experience "signals" scrambling combined with an inability to use GPS or receive cell service in certain areas, I roll my eyes over the possibility (but understanding) that some form of this signal jamming is easy to do and may likely be true, especially if targeted on a building or group.

The "pushing out" effect refers to HAARP's ability to temporarily heat and excite small areas of the ionosphere, creating *artificial disturbances* for study. This heating does cause slight local expansions in the ionosphere, though these effects are said to be temporary and localized. Some scientists call bullshit and claim that massive discharges came back to Earth at volts one-hundred times greater than a lightning bolt. There are no standards set for this because once again, a few humans seem to have little understanding about how life and Earth are symbiotic and interact with both the cosmos and quantum particles. Maybe it's too late and we've already irrevocably disrupted migratory patterns and minds.

HAARP generates electromagnetic fields during its operation, primarily through radio frequency emissions in the 2.8-10 MHz range, local heating effects in targeted ionospheric regions, and induced electromagnetic waves. Studying these fields provides us with an advanced understanding of ionospheric physics, improved communications technology, better space weather prediction, enhanced knowledge of aurora formation, and development of new radio technologies.

However, there are concerns ranging from local environmental impacts from facility operation and electromagnetic interference with nearby systems to energy consumption and secret military applications.

As I consider future sci-fi books, I think more positively about how HAARP might be used for extraterrestrial communication, using focused ionospheric modulation to create coherent signals visible from space.

This could lead otherworldly advanced beings straight to and through our portal systems. Temporary "windows" in the ionosphere might be opened for enhanced deep space transmission to contact life on other planets. We might also generate complex plasma patterns as visual signals or for mass decoy purposes to cloak our planet, making it invisible to Darth Vader types—or if they did get through, we could send bots in to battle them and then clean up the holes in our ozone.

Unfortunately, a highly advanced interplanetary civilization (or even advanced humans on Earth) would likely use the ionosphere to disrupt global communications during an alien invasion or create localized electromagnetic disturbances to make people think that certain mythic human-hybrids or Gods exist.

The reality is that HAARP has contributed significant scientific advancement to our understanding of the ionosphere and radio communications. And its potential applications continue to inspire both scientific and creative speculation.

Indeed, beaming high-frequency radio waves into the ionosphere, a practice that heats it up, has positive and negative implications too, including understanding how solar storms might impact or disrupt our satellite systems. It's impact on people depends on the scale, intensity, and intent of the technology.

Critics rightly argue that large-scale manipulation of natural complex systems, like the ionosphere, might have consequences that scientists and their respective governments cannot fully predict under a deeply compartmentalized *need-to-know* air-tight system. This sort of secret *Schrodinger's box* raises questions about whether deeply hidden unacknowledged special access projects should engage in experiments that unknowingly impact the planet and humanity without humanity's informed consent or robust oversight.

Likewise, any futuristic sort of ionospheric applications requires massive technological advances and a much deeper understanding of ionospheric physics than we allegedly currently possess. However, the HAARP facility remains an important *allegedly* private university-led scientific research installation, helping to better understand our atmosphere and its interaction with space weather and radio communications.

Similarly, since HAARP is pushed upon the public as completely harmless and we are researching quantum effects and its relationship to both science and the sacred, let's try to skin this ionospheric cat.

PART IV

Beyond Conventional Physics

CHAPTER FOURTEEN

The Atmosphere
A Cosmic Quantum Laboratory

Considering that the ionosphere is a vibrant region of Earth's upper atmosphere, where particles dance under the influence of both classical and quantum mechanics, think of it as a nearly perfect cosmic laboratory: It's teeming with charged particles—electrons and ions—that behave in ways which challenge human knowledge and intuition.

The ionosphere is where the mysterious realm of quantum plasma physics comes into play. For example, electrons do not follow predictable paths. Instead, their behavior adheres to quantum rules, like wave-particle duality, where particles act like waves and vice versa. This duality significantly influences how radio waves—used for communication and navigation—interact with the particles in the ionosphere.

Imagine a radio wave rippling through a crowd of particles that can either absorb, scatter, or amplify the signal based on quantum interactions. As energy levels rise, these quantum effects grow more pronounced, shaping the ionosphere's behavior in ways that defy the classical models we use to describe nature. It's an electrifying blend of science and the sacred, where the supernatural (science we don't yet understand) meets the edge of our atmosphere.

Additionally, the ionosphere is also a playground for quantum tunneling—a mind-bending phenomenon where electrons defy expectations and pass through energy barriers that, classically, they shouldn't be able to cross. Picture an electron (a flavor of neutrino) approaching a wall. It doesn't bounce back but "teleports" to the other side. This is quantum tunneling in action, a process that profoundly influences the ionosphere's behavior.

Subsequently, tunneling affects ionization rates—the process of atoms losing or gaining electrons—and the transport of electrons across energy barriers. These interactions are so much more than theoretical curiosities; they play a critical role in the ionosphere's dynamics, particularly in how energy flows and particles behave.

Instruments like HAARP, which interact with the ionosphere to study and manipulate its properties, must carefully account for this quantum tunneling and its effects. We are essentially trying to conduct an orchestra from Earth where some of the instruments of the ionosphere play by the rules of another dimension. This certainly adds highly delicate and detailed complexity to the symphony of interactions in our planet's upper atmosphere.

It is important to remember that Earth's ionosphere is alive with an ancient pre-human dance of quantum energy level transitions, where atoms and molecules absorb or release discrete packets of potential, known as quanta (the smallest discrete units of energy, matter, charge, or angular momentum). When HAARP's powerful radio waves ripple through the ionosphere, they excite both atoms and molecules, nudging electrons to higher energy levels or letting them cascade back down. Unfortunately, no matter what we're told, these transitions have real-world effects. Remember, when it comes to airplane crashes, nuclear bombs, lost missiles, uninformed consent experiments, lab-leaked viruses, or other sorts of intended or unintended *accidents*, oops doesn't cut it…but severe plausible denial and copious coverups rule. Asses will be covered before shields to the public are provided.

Light emission, such as the auroras that shimmer across our polar skies, is a direct eye-catching, mesmerizing result of such quantum leaps. Energy absorption heats the ionosphere, altering its density and behavior, while particle interactions ripple outward, influencing everything from satellite communications to navigation systems. It is indeed a cosmic symphony, where HAARP tries to act as both conductor and participant, unlocking the quantum secrets of Earth's upper atmosphere. Whether or not it does so without harm to humans or other life on this planet remains to be seen—or not.

Quantum Plasma Effects of the Ionosphere

The ionosphere is more than a sea of charged particles—it's a collection of quantum behaviors that reveal deep interconnectedness, like Bohm's folding and enfolding in his implicate and explicate order made manifest. Picture it as a vast invisible wave of electrons, each one moving in concert with countless others under the nonsensical anomalies of quantum mechanics. Collectively, these phenomena are what we call *quantum plasma effects*.

HAARP deep-dives into this quantum-rich environment, exploring plasma waves. Another phenomenon known as *collective electron oscillations*, seems to sway 'to and fro' as a crowd, propelled by electromagnetic forces. Even more fascinating are *quantum coherent states*, where particles act so synchronously that they behave as if they are a single entity.

Each particle, like fish or birds swimming or flying in synchronicity, somehow "knows" what the others are doing. Oscillations and coherence shape how the ionosphere responds to radio waves, solar storms, and even human-made influences like HAARP's transmissions. It's like a delicate harmonization of high-wireless energy where quantum rules govern it all—and it provides an exciting look into the universe's fundamental principles at work right above our heads.

And because I can't resist the potential for sci-fi consideration, let's explore some creative fiction-future research directions for HAARPs:

Quantum-based Diagnostic Tools

Premise: In a future where humanity colonizes planets with hostile atmospheres, quantum diagnostic tools allow real-time analysis of environmental conditions. These devices 'visualize' and precisely measure the molecular structure of air, detect hazards, and map alien pathogens.

Plot Twist: A planetary colonization mission discovers that their diagnostic tools detect atmospheric changes *before* they occur, hinting at a mysterious interaction between quantum systems and time itself that resembles something like human déjà vu. Could the planet be "aware" and intelligently adjust itself accordingly, even if inaccurately, to anything it deems a threat?

Quantum Computing for Data Analysis

Premise: A vast quantum computer network called the X-Grid processes exabytes of data to maintain Earth's environmental balance. However, it starts uncovering irregular "patterns" in the ionosphere that suggest interference from an advanced cosmic intelligence.

Plot Twist: The interference, it turns out, isn't extraterrestrial—it's previously created quantum AI from the future, sending a warning back in time to humans unknowingly conducting catastrophic experiments.

Quantum Simulation of Ionospheric Processes

Premise: Scientists use quantum simulations to predict massive ionospheric storms, which devastate Earth's communication networks. The simulations reveal mysterious ionospheric "waves" behaving as though controlled by an unknown entity.

Plot Twist: The simulations unlock a pathway to communicate with an ancient, sentient lifeform dwelling in the ionosphere, raising ethical questions about how to coexist with interdimensional beings' humans never imagined.

Macroscopic Quantum Coherence

Premise: Quantum coherence at macroscopic levels leads to breakthroughs in "quantum plasma shields," enabling instantaneous deep-space travel.

Plot Twist: During one voyage to another galaxy, the shield develops coherence with a parallel universe, revealing alternate versions of the human space travelers' lives—forcing them to collectively choose between universes when the shield begins to destabilize. The choice inevitably erases everything meaningful one of them holds dear —but they won't know which person it impacts until *after* the choice is made and the leap concludes.

Quantum Phase Transitions in Plasma

Premise: Experiments with quantum phase transitions in plasma reveal exotic states of matter that can freely power the entire planet. However, one experiment creates a plasma state that behaves like a sentient organism.

Plot Twist: The plasma entity communicates cryptic warnings to the researchers and to unsuspecting Earthlings through dreams about catastrophic future events. It strongly hints that humans have the power to birth an irreversible "quantum apocalypse" or prevent one depending on how they exercise collective consciousness.

Quantum Chaos in Ionospheric Systems

Premise: Scientists investigating quantum chaos in the ionosphere discover patterns resembling neural brain activity.

Plot Twist: Here, you decide the plot twist.

Each sci-fi speculation opens a portal into narratives that spin advanced quantum technology onto the human condition, our ethical dilemmas, and the consequences such adventure brings when we explore and manipulate unknowns. While it is a relief that ethics courses are mandated for scientists at both the graduate and post graduate level, especially as it pertains to research—here's to hoping we don't collapse the entire wave function because a few particle scientists or their science-ignorant leaders misbehaved.

Linear Force Lines

Linear force lines or field lines are *imaginary* lines that show the direction and pattern of electric or magnetic forces acting in a physical system. They are an essential *conceptual* tool in studying magnetic and electric fields because they help visualize the strength and direction of these fields in space.

Understanding force lines has practical scientific applications and implications for both advanced technologies and speculative sci-fi. Humans use them to simplify complex systems, for experimental insight, or to design mechanical or electrical equipment. One property of linear force lines is that they never cross (except at points of equilibrium), since a force can only act in one direction at any given point. The density or spacing between lines indicates field strength—closer spacing means stronger forces.

Comparatively, in nuclear fusion reactors like tokamaks (devices that use powerful magnetic fields to confine hot plasma), the magnetic field lines are used to confine plasma. Mastery of force lines could revolutionize controlled fusion, which would bring us nearly limitless energy.

Quantum Field Manipulation

Force lines have, I suspect, applications in quantum field theory or the manipulation of quantum systems. For instance, in quantum computing, tailored electric or magnetic fields might be able to one day manipulate qubits with high precision. Likewise, advanced magnetic force lines could create stable quantum traps for particles or materials enabling levitation re: lifting or lasering heavy stones in architecture.

Force Line Engineering

Speculative sci-fi scenarios might involve engineering "programmable" force fields for energy shielding where force lines might be manipulated to repel physical or energy-based attacks.

Additionally, propulsion systems could be used to control magnetic or electric fields to create spacecraft propulsion (e.g., magnetic sails). At the quantum level, vacuum energy fields often interact with the vacuum of space. Force lines might be used to manipulate Casimir forces or tap into vacuum energy for advanced propulsion or energy generation.

Quantum Entanglement Networks

Structured electric or magnetic fields might be used to influence quantum entanglement on macroscopic scales, potentially forming the basis of faster-than-light communication, travel, or even a quantum internet.

Additionally, linear force lines are more than just a visualization tool for engineers, artists and sages; they bridge theoretical physics with practical scientific applications using *imagination*, like writers, sages, painters, poets, and musicians do.

As we deepen our understanding of the quantum world, force lines are clues that enable the manipulation of fundamental forces in ways that seem unthinkable today but form the foundation of future technologies and become some of the greatest stories ever told (or sold). From controlling energy, to enabling sci-fi shields or fields, propellant-less propulsion, and quantum networks, force lines buzz with a lot of unzapped potential.

CHAPTER FIFTEEN

Through the Wormhole and Beyond
How to Lose Your Mind in Space-Time

"*What happens when you get to the end of things?*"
A physicist named John Wheeler spent his life and career obsessed with this question and eventually arrived at the conclusion that space-time is not the true fabric of the universe. Like Shamans and scientists before him, Wheeler was on a hunt for a deeper reality beneath, between, or around space-time. He pondered if time and space were linear like the arrow or more of an infinity symbol that keeps looping back to us.

Time, like consciousness, is difficult for humans to precisely define or locate, even though each of us claim with relative certainty and little reflection that we absolutely know and understand time.

Wheeler, who studied under Niels Bohr and held conversations with Einstein, broke the quantum space in particle, nuclear, and gravitational physics. During his career, he asked questions about matter that remind me of Bohm—that perhaps matter is really all waves or one wave of gravity folding itself into compact spheres that resemble subatomic particles on the outside, but on the inside are empty. He called these compact spheres "geons". He asked other questions: *Can particles be reduced to geometry or pre-geometry?* And *Is space-time twisted and warped with holes and handles wrapping around itself?*

Maybe electric fields, he thought, could burrow and hide in holes to re-emerge someplace else—negatively charged when they disappeared—and positively charged upon return. Wheeler gave this twisty licorice pondering of space-time a couple of cool-sounding names—wormholes and quantum foam—and eventually decided upon a loop cycle, one void of gravitational disappearing acts.

Wheeler claimed that wormholes are pervasive but unstable and collapse, a kink where the walls cave in and the wormhole vanishes causing singularities within their remains. He surmised that if space-time vanishes in the blink of an eye, then *something* lies somewhere, beyond, beneath, or between. There just can't be nothing! Fortunately, thanks to help from NASA (again), Wheeler came up with a name that stuck: Black holes—the space where reality and time end.

However, wherever he went with space-time, *something* was always left—the gnawing knowing that quantum mechanics destroys the concept that everything "sits out there" away from us. It was an awakening that Bohm also, to some extent, and Shannon, to a greater one, already realized: all things great and small are interconnected, and comprised not of matter, but of information, a dimension free character of bytes and bits, 0's and 1's—binary choices or qubits. Good and evil. Yin and yang. Darkness and light. Schrödinger's cat and Pavlov's dog. Dualism seems to be a necessary human condition in a 3-4D world.

But, our world, Wheeler concluded, is a universe of choices—not matter, not space-time—but choices leading back to the *participator* (rather than the observer), one looking into an endless row of mirrors and seeing his or her reflection, ad infinitum. If there is no observer, there is not God, because God is us. This is an extraordinarily difficult concept for the faithful, especially those who like to separate God from humans, to ever accept—unless space-time is a fiction too, useful in a 3-4D world but not really needed.

In Wheeler's world, it is us who decide which slit a photon, sent from a quasar, passes through—because its already happened. In other words, we are a holographic fractal caught in the matrix of a loop, living within a seed or cell called the universe. Humans are receivers, containers, and transmitters of information and energy—the ultimate secret code, passed from one generation to the next through DNA—and we are an interdimensional field—the bridge between the classical and the quantum.

Wheeler's ponderings led to "it from bit", like Shannon, and brought us to the age of quantum computing and teleportation, black hole entropy, and the holographic principle of quantum gravity. What troubled Wheeler, according to his journals held at the American Philosophical Society in Philadelphia, is how all of us, hammering away at our individual realities, produces a collective one—one that never completely agrees, and often feels disjointed and chaotic.

Unfortunately, time as humans know it ran out for Wheeler when he transitioned from this planet. One of his last journal entries at the age of ninety-five asks: *"Hope produces space and time?"* He was onto solving a great mystery with such a profound question.

Today, some physicists propose that spacetime is not a fundamental backdrop of the universe but an emergent phenomenon—rising from the interactions of deeper quantum constituents. This concept, often referred to as *modular spacetime*, reimagines spacetime not as a pre-existing fabric but as something defined by the relationships between objects.

Intriguingly, recent observations from the Dark Energy Spectroscopic Instrument (DESI) hint at phenomena consistent with Wheller's view—perhaps even edging toward predictions associated with string theory. The nature of time, as Wheeler shows us, is one of the most captivating questions in physics, philosophy, Sci-Fi, spirituality, and human experience.

In modern physics (particularly since the days of Einstein's theories of relativity), time is understood as a dimension interwoven with space in a four-dimensional "spacetime" continuum. Unlike our everyday experience of time flowing uniformly and in linear fashion, relativity shows that time is relative—it can dilate or contract depending on motion and gravitational fields. Einstein taught us that two observers moving at different velocities or in different gravitational fields will experience time differently.

A crucial aspect of time is its apparent directionality or "arrow". While most physical laws are time-symmetric (they work the same forward and backward), the Second Law of Thermodynamics introduces asymmetry through entropy. Systems naturally tend toward higher entropy states, giving time its humanly familiar and commonsensical forward moving direction. This explains why we remember the past but not the future, why a broken glass won't reassemble itself and why a human being can't spontaneously age backward to become decades younger.

Conversely, in classical physics, "vertical time" refers to the time it takes for an object to travel vertically—either up or down—under the influence of gravitational forces. However, there is a much deeper interpretation of vertical time when we step beyond classical models.

A.R. Bordon, for example, proposed that vertical time represents a dimensional axis distinct from linear time, allowing consciousness to access nonlocal information streams. Through experiments, Bordon suggested that by tuning *neural frequencies*, human beings could navigate vertical timelines, accessing pasts, presents, and futures that coexist simultaneously—not merely to simply *observe* them but to actively *change* them.

Similarly, we encounter the concept of *sidereal time*, where growing evidence suggests psychic phenomena may be influenced by the Earth's geomagnetic environment and its alignment with the broader cosmos.

Researcher James Spottiswoode uncovered significant findings showing that remote viewing (aka neurosensing or 'dimensional peering') success rates varied according to sidereal time—specifically linked to Earth's orientation relative to the center of the Milky Way Galaxy. He worked with SRI International and SAIC (Science Applications International Corporation), analyzing remote viewing data under government-funded programs. His contributions were methodological—focusing on effect size, statistical rigor, and experimental design.

Published in the *Journal of Scientific Exploration* in 1997, Spottiswoode's Cognitive Sciences Laboratory research in Palo Alto, California indicated that shielding subjects from magnetic field variations, and accounting for sidereal alignment, measurably increased anomalous cognition results akin to Kozyrev mirror research. These findings support the idea that both geomagnetic forces and cosmic orientation likely play critical roles in shaping human extrasensory perception.

Additionally, subjective experience of time differs markedly from physical time. We experience time as flowing in a future direction, with a distinct past, present, and future. We can only live in the present but many of us remain unaware or oblivious of the *Now*, being firmly future-oriented—the grand illusion. But what and where exactly, is *Now*? And how long does *Now* last? Is *Now* a lifetime, a second, a flash, or a blip?

Regarding the flow of time, some philosophers, and even some physicists, argue that this "flow" we call time is an illusion—that all moments exist simultaneously in a "block universe," with our consciousness creating the *sensation* of passing time. Our perception of time's speed also varies greatly based on our mental state, age, and activities. The older humans become; the faster time *seems* to pass. The perception of time speeding up as we age is a fascinating psychological and neurological phenomenon. We use time to measure change within our existence in 3-4D space.

For example, when you're six years old, one year is almost seventeen percent of your entire life—so it feels significant. By the time you're fifty, one year becomes two-percent of your life, making it *feel* comparatively shorter. Likewise, the brain encodes time based on experiential richness.

In childhood, the world is new, and our existence expands at a frenzied pace as we proceed into many "firsts" (e.g., first day of school, first bike ride, first crush, first concert, first drink, first fuck). This builds a foundation of densely packed memories, good, bad, and ugly. We maneuver through childhood and our teenage years, feeling as if we can't grow up soon enough. We use play, direction, programming, and observation as tools to move mostly linearly through time.

And as we age into fully-fledged adults, and these *firsts* morph mundane, life typically becomes blasé or routine. Everything becomes familiar unless we try to *spice things up*. Fewer novel experiences mean fewer dense memory clusters, lending credence to the secondary illusion that time seems, in hindsight, to pass faster than light-speed. Age-related changes in dopamine levels and brain processing also influence our perception of time. Slower processing makes external events appear to move faster relative to our internal clock.

Quantum Physics Perspective of Time

Quantum mechanics doesn't directly explain our subjective perception of time, but certain interpretations of quantum theory allow for sci-fi-like speculative connections.

In quantum mechanics, time isn't absolute. It's often treated as a parameter (not a measurable observable like position or momentum). This echoes Einstein's relativity, where time depends on the observer's frame of reference. Aging could metaphorically reflect a "shift" in how we experience the passage of time, though admittedly, this isn't grounded in quantum physics.

Another example…if we could travel to the Andromeda galaxy and back at faster than light speed, millions of years on Earth would have passed and everything we think we *owned* and everyone we knew and loved (or loathed) would be long gone—maybe even the planet too.

But let's look at quantum entanglement, specifically where it suggests non-locality—where changes to one particle affect another instantaneously, seemingly bypassing time as we understand it. This hints that time might not be fundamental, but *emergent*. If time emerges differently at varying "scales" (biological, psychological, or even universal), such factors might explain why our subjective perception of time changes.

Likewise, time's one-way 3-4D flow is tied to the increase of entropy (disorder) in the universe, as per the Second Law of Thermodynamics. As we age, we perceive time differently because our brains, which operate as entropy-driven systems, encode information less effectively. This resonates with quantum thermodynamics perspectives on the universe's unfolding—and perhaps, on some level, mirrors Bohm.

Remember, in quantum mechanics, the act of observation collapses probabilities into a definite state. Similarly, our perception of time might be influenced by our "observing" and participating in universal consciousness. Over time, the brain's ever-changing and morphing cognitive and memory systems could also parallel this collapsing process, altering our *perception* of time's flow.

Interestingly, quantum physics fundamentally challenges our classical understanding of time, suggesting it may not be as linear or universal as we experience. While this doesn't directly explain the psychological phenomenon, it underscores that time is a deeply subjective and relative construct—like personalized *perception*—and not just in physics, but so too, in our minds.

The debate about time's fundamental nature continues. Is time a basic feature of reality, or does it emerge from more fundamental phenomena? Does time flow continuously or come in discrete units at the quantum level? Could time have begun with the Big Bang, or does it extend infinitely? These questions intersect physics, philosophy, and our deepest intuitions about existence.

Furthermore, past-future causation refers to the typical understanding of causality where events in the past influence or determine events in the future, but not vice versa. This aligns with our everyday experience of time's arrow and is fundamental to most scientific and philosophical frameworks. This form of causation is characterized by:

1. Temporal order: The cause must precede the effect in time. For example, a ball rolling down a hill happens after it was pushed.

2. Unidirectional flow: Information and influence flow from past to future, not backwards. We can affect tomorrow's weather by releasing greenhouse gases today, but tomorrow's weather cannot affect today's emissions.

3. Asymmetry: While we can use past events to predict future ones (like using weather patterns to forecast storms), we generally cannot use future events to retroactively influence the past.

Asymmetry becomes particularly interesting in discussions of quantum mechanics, where some interpretations suggest more complex forms of causation like determinism and free will, chains of cause and effect, or methodology, which relies heavily on past events predicting future outcomes.

Wormholes

A wormhole is a theoretical structure in space-time that essentially creates a tunneled shortcut through space and time, connecting distant points in the universe. It's a bit like looping a hose at each end to considerably close the distance between two (or more) places.

Wormholes, if we could find and use them, interact with time in substantial ways. Because they distort spacetime, traveling through one potentially allows movement not just through space but also time. This would likely enable travel to different time periods, though this raises complex and interesting paradoxes that physicists and philosophers, and many scientists, still debate.

Creating a wormhole in a lab is currently beyond public domain science and technological capabilities. While some physicists have proposed methods involving exotic matter with negative energy density, only clandestine labs have found a practical way to generate and manipulate such matter. While we're told that the energy require-ments would be astronomical, and maintaining stability would be extremely challenging, it has allegedly been done—but people are understandably too nervous to discuss it without clearance.

Wormholes are related to black holes (which may or may not exist, but if they do, come in a multitude of sizes and may be found or created anywhere) in that both involve extreme warping of spacetime. But they are distinct phenomena. While black holes have a one-way event horizon, from which nothing can escape (this is a fiction), a traversable wormhole allows two-way travel.

When humans travel through wormholes, it allegedly leads to distant parts of our own or other universes—allowing us to reach galaxies or alternate dimensions that would otherwise be impossible to visit. A wormhole could also open a window to unknown physical constants leading to radically different forms of matter or alternative evolutionary paths where life developed under completely different conditions. If we zoomed through a wormhole, we might encounter civilizations that evolved in higher or lower dimensions, or consciously sentient beings made of exotic forms of matter.

Life forms in these places would likely be utterly alien to human understanding, sometimes existing as patterns of energy, collec-tions of quantum information, or forms of consciousness that don't require physical bodies as we know them. I've read that making such information public would likely lead to societal collapse and a sort of collective existential crisis from which humans might not recover.

Allegedly, these otherworldly or interdimensional life forms perceive reality in ways human beings can't imagine, like how two-dimensional beings would be completely terrified and unable to comprehend our three-dimensional existence.

Consequently, it's important to note that wormholes are mathematically possible according to Einstein's equations, and we do have *observational* evidence they exist.

In 2022, physicists at the Delft University of Technology created a quantum system that simulated properties of a traversable wormhole using a quantum computer. This wasn't a real space-time wormhole, but rather a quantum simulation that exhibited similar mathematical properties. The experiment used quantum bits to simulate how information might travel through a wormhole-like connection.

Similarly, scientists at Google, in collaboration with other institutions, announced in December 2022 that they had created a "baby wormhole" using their Sycamore quantum processor.

Again, it's crucial to understand this was also a simulation of wormhole dynamics rather than an actual space-time wormhole. Yet these important experiments demonstrate how quantum information behaves in ways that mathematically mirror how matter *might* move through a wormhole.

Another significant development is in the study of "analog black holes" or "sonic black holes," where scientists create conditions in laboratories that mimic gravitational phenomena. While not wormholes per se, they help scientists understand some of the physics involved in wormhole formation.

Astronomically, scientists have been studying Einstein-Rosen bridges (the technical term for certain types of wormholes) through their theoretical connection to quantum entanglement. Some physicists propose that entangled particles might be connected through microscopic wormholes, though this remains highly theoretical in the public domain.

The primary challenge in creating an actual wormhole is that it requires exotic matter with negative energy density, an enormous amount of energy (found in miniscule amounts of space), technology to stabilize the wormhole against collapse, and effecting shielding to prevent lethal radiation for anything passing through.

Currently, simulations and theoretical work help humans better understand the quantum physics—to a small degree. Yet the experiments are groundbreaking in terms of quantum leaps even if they are mathematical models that behave like wormholes rather than creating actual shortcuts through space-time.

Notably, quantum theory also says information can't be destroyed, but general relativity suggests that anything falling into a black hole is lost forever. This paradox has sparked a lot of ideas involving entanglement, like the ER (Einstein-Rosen Bridge) =EPR (Einstein-Podolsky-Rosen paradox; quantum entanglement) hypothesis, which suggests that entangled particles might be connected by tiny wormholes—or that they are the exact same thing at a fundamental level. Perhaps information is stored on the fringes of a black hole (if they truly exist), rather than lost or destroyed.

Project Looking Glass

"Project Looking Glass" is a term that appears in various truth crusader theories and speculative claims, often linked to secret government projects involving time travel, advanced technology, or psychic abilities. The concept lacks credible evidence and primarily exists in the realm of internet speculation, alternative media, and fiction—but there's more to it than this.

Governments have spent a lot of taxpayer dollars on time conundrums and have sought to time-jump, harness time, or otherwise predict or control events in time. Here's an overview of the claims associated with Project Looking Glass:

Time Travel and Advanced Technology

- Project Looking Glass is often described as a secret program allegedly run by the U.S. government (mostly shadowy organizations deeply buried within clandestine operations) to develop technology capable of looking into the past and future.

- Proponents claim the project involved three advanced devices (a sphere, a laser, and a cube) that could bend time and space, allowing users to see potential timelines or predict future events.

- This was the rudimentary experiment that led to improvements in neurosensing (combining technology with biology), to advance vertical timeline manipulation.

Connection to Alien Technology

- Some theories suggest the technology was reverse-engineered from extraterrestrial artifacts or derived from suppressed scientific breakthroughs gleaned from alien or foreign adversary technology.

Montauk and Philadelphia Project Links

- *Project Looking Glass* is sometimes tied to other projects in legends like the Montauk Project or the Philadelphia Experiment, each involving time travel, invisibility, or multidimensional experiments. The projects are said to have involved the US military and private contractors conducting experiments in teleportation, mind control, and other areas. Opponents of this legend claim that the entire story is fictional bullshit and was fabricated by two men, Preston Nichols and Stewart Swerdlow, who were perhaps hoping to become famous.

  Alternatively, it is important to note that a few US-based *Project Looking Glass* or *Operation Looking Glass* exercises did exist, mostly to do with military command, control, and communications for launching ballistic missiles or monitoring the globe for nuclear attack.

- **Predictive Capabilities**: A recurring claim is that even the legitimate projects were used to foresee significant global events or to manipulate timelines to ensure desired outcomes.

While the claims about Project Looking Glass remain speculative, they often touch on actual scientific theories like space-time and Einstein's Theory of General Relativity, which describes how massive objects warp space-time. Einstein's theory of GR is the foundation for concepts like wormholes and time dilation. Remember, wormholes, theoretically at least, connect distant points in space-time and might allow for faster-than-light travel or time travel.

Conspiracies suggest that technology like Project Looking Glass manipulates space-time at will, but there's no declassified experimental evidence supporting the creation or control of wormholes. Even if they exist, maintaining stability (to avoid collapse) is a major hurdle in public or Earth-bound labs.

As previously discussed, quantum theory explores phenomena like particles existing in multiple states simultaneously or being "entangled" across vast distances. This idea is also often linked to ideas about alternate timelines or parallel realities. Yet these types of theories purposely exaggerate quantum concepts for conceptual purposes, proposing that machines can "observe" or "select" alternate timelines. While quantum mechanics is real, scaling these effects up to macro-level events or human perception isn't currently advised or feasible.

Remember too, that the Butterfly Effect in chaos theory, where small changes in initial conditions can drastically affect outcomes, make accurate predictions of complex systems (like the weather, or human history) nearly impossible for extended periods—except perhaps in block or vertical time.

Likewise, any claims about "seeing the future" ignore the unpredictability of uncontrollable, chaotic systems. Even if someone had detailed knowledge of initial conditions, predicting human choices and interactions remains incredibly difficult and complex.

Some interpretations of quantum mechanics (like the Many-Worlds Interpretation) suggest that every quantum decision creates a branching timeline, resulting in infinite parallel universes—and this may influence or handicap our ability to accurately and precisely predict the future under real-world conditions.

Likewise, Project Looking Glass theories assume that technology somehow views or navigates these multiple universes. However, it's important to remember that the multiverse remains a speculative idea (although it might be the best candidate we have), and there's no evidence we can *consciously* interact with alternate universes, let alone "choose" timelines.

Subconscious, internal interaction with alternate dimensions is another quantum matter altogether—possible I suspect, through lucid dreaming or dimensional peering. If brain microtubules and DNA are biological links or keys to quantum gates possessed of supernatural capabilities—the technology we develop to aid mind and body in such endeavors requires the sentient awareness of a participator to work.

Alternatively, reverse-engineering alien technology is a recurring theme we've heard in Congressional hearings, read in books, or heard on podcasts and documentaries. All claim breakthroughs like faster-than-light travel, extraterrestrial visitors, strange phenomena or out-of-this-world manipulation of matter. While there are ongoing studies in areas like fusion power, anti-gravity research, and AI, there's also hidden evidence that these fields involve extraterrestrial artifacts.

Cultural Influences Regarding Time

Stories like H.G. Wells' *The Time Machine* (1895) and blockbuster films such as *Back to the Future* have long fueled humanity's fascination with time travel. From *Star Trek* to *Stargate*, *Star Wars* to *Dune*, science fiction has consistently explored wormholes, alternate dimensions, and the manipulation of time. These imaginative narratives don't just entertain—they often inspire real-world scientific inquiry and research.

Concepts once relegated to fiction, like those in the speculative Project Looking Glass, have increasingly informed experimental technologies and theoretical models, bridging the gap between imagination and innovation. For example, the Alcubierre Drive – A theoretical propulsion system inspired by the warp drives of *Star Trek*, is now studied seriously in the context of general relativity and spacetime manipulation.

Additionally, nonlocality and quantum entanglement were once thought purely fantastical. These ideas—echoed in films like *Interstellar*—are now the foundation of quantum communication and cryptographic technologies.

Moreover, the public now uses technology once the realm of science-fiction. Devices like iPhones and Androids were once viewed as improbable futuristic devices in *Star Trek*. Now, they connect us instantly across the planet, act as portable computers, and use GPS satellites—once a military-only system, practically perfected in 1984, and rooted in Cold War-era space science.

Voice Assistants and AI (e.g., Siri, Alexa, ChatGPT) echo the talking computers and intelligent AIs of science fiction, enabling real-time speech interaction, translation, and decision-making while video conferencing and holograms, once featured in *Star Wars* or *Blade Runner*, are now known as Zoom or FaceTime. Holographic projections and deep fakes are also emerging in events, videos, and advertising.

Smartwatches and AR glasses (like Apple Vision Pro or Meta's Ray-Bans) resemble gear from *Minority Report* or *Iron Man*, combining real-world interaction with digital overlays and immersive environments, once only dreamed up in stories like *Ready Player One*, are now accessible for education, warfare training, gaming, and even medical simulations.

The point is most people don't learn about clandestine research until its technology hits the market years later or provides a benefit. The mistakes and losses are usually buried and/or denied.

During the Cold War, secretive government projects like MKUltra (mind control experiments) and Operation Paperclip (recruiting German scientists) bred public distrust but also led to medical and nuclear advances.

This sort of deep-black history fuels speculation and ethics discussions about what other "hidden" technologies, labs, and experimental knowledge governments possess—such as epigenetic labs cloning humans or creating human-hybrid-like chimeras (imagine a half-fish, half-human or any other mythological character brought to life). Echoing Democritus, we have the technology to do this, but should we? Unfortunately, I feel it is a temptation and a line that some scientists in secret labs crossed long ago.

Projects like the Manhattan Project (atomic bomb) and stealth technology programs (e.g., the F-117 Nighthawk) were highly classified, proving that governments can and do keep secrets. This lends credibility to the idea that something like *Project Looking Glass* might exist, even though there's no declassified evidence that it does.

As a species, human beings are naturally drawn to stories that provide secret knowledge or explain complex phenomena in simple ways. Good Sci-Fi and stories such as *Project Looking Glass* play into this desire, offering an alluring mix of mystery, power, and rebellion against nature, mortality, and authority.

Notably, Dan Burisch (aka Dan Crain) was a biowarfare microbiologist who allegedly had a Q Clearance and worked at S-4 at Area 51 like Bob Lazar, in addition to his roles at the Defense Intelligence Agency and the Naval Research Laboratory.

Burisch is a well-known figure linked to one wild *Project Looking Glass*. He claimed that the U.S. government developed a prototype device based on alien technology that allowed them to view "probable" futures by manipulating space-time. He said he was tasked to study another ancient alien device and that Sumerian seals were part of the key to unlocking timeline secrets. It's possible that Burisch found himself eighty-three miles northeast in Chaco Canyon studying paradoxical artifacts the likes of A.R. Bordon and the folks at the NSA's Advanced Intelligence Contact Organization (not its real name).

Once the looking glass device was working, all dates seemed to converge on December 21, 2012, as being the end of the world. Since we're still here, I can only assume that the entire story was fictional, part of a psyops, or that our black ops and unacknowledged special access projects should be credited for successfully moving Earth's timeline to avert apocalypse.

According to Burisch, *Project Looking Glass* was discontinued because it revealed future scenarios that were "too dangerous" or "paradoxical," such as societal collapse or nuclear war. "I was always told that monsters weren't real," he said during a recent YouTube interview on *Nightshift*. "And then I find out that they are, but they live in the heart of men." A fair warning for our future if there ever was one.

Notably, researchers uncovered Burisch's accolades, awards, achievements and degrees. It appears he wasn't lying about his high-level, credible work even though he was ruthlessly castigated years ago by some of the biggest names in ufology. I encourage anyone to watch Burisch's interviews, conduct their own research, and draw independent conclusions.

Descriptions of a Project Looking Glass Device (It is alleged there is more than one)

Many anecdotes focus on the device itself, sometimes described in highly specific yet unverifiable terms:

Design:

- The device is often depicted as a spherical or circular structure, sometimes involving liquid crystals, electromagnetic fields, or plasma.
- It is said to create a "looking glass" effect, projecting images of possible futures or alternate realities on a screen or into the minds of operators.

Functionality:

- Operators would input variables or "questions" into the device, and it would generate visions of potential outcomes.

- These visions were not deterministic but probabilistic, showing likely future events depending on current circumstances.

Claims of Timeline Manipulation

- Some accounts suggest that the device revealed a point in the near future (sometimes referred to as the "singularity" or "timeline convergence") where all possible timelines converge into one.
- This convergence is often portrayed as inevitable, leading to a transformative event for humanity, such as enlightenment or global upheaval.

Avoiding Catastrophe

- The project allegedly sought to prevent catastrophic events, such as world wars or environmental collapse, by steering humanity toward preferable outcomes.
- According to some accounts, this manipulation caused unintended side effects, including "timeline instability."

Connection to Extraterrestrials

Many anecdotes link *Project Looking Glass* to aliens and UFOs:
- The technology is said to have been derived from alien artifacts recovered from Roswell or other UFO crash sites.
- Some accounts claim that extraterrestrial beings directly shared this knowledge with government officials.

Warnings from Aliens

- Anecdotal claims suggest that aliens warned humans against using the device irresponsibly, as it could destabilize the fabric of space-time.

Alleged Use in Global Events

Project Looking Glass is sometimes tied to major historical or speculative events:

- Some conspiracy theorists claim the device was used to foresee or influence events like the September 11 attacks, wars, or elections.
- There are unverified claims that world leaders and elites used the technology to consolidate power by "peeking" at future events.

QAnon and 2020's Predictions

- In recent years, *Project Looking Glass* became associated with QAnon conspiracy theories, where it was claimed the device revealed a "great awakening" or coming upheaval.
- Proponents claimed the project could expose corruption or lead to a utopian future, depending on humanity's collective choices—a bit like hordes of people concentrating on random number generators to slant a particular outcome one way or another.

Alleged Discontinuation

Anecdotal accounts frequently mention that *Project Looking Glass* was discontinued for various reasons:

- Using the device supposedly revealed paradoxical or destabilizing effects, such as creating alternate realities or timelines that conflicted with each other.
- Operators allegedly experienced mental breakdowns or other side effects when exposed to the device's visions.

Ethical Concerns

- Some anecdotes claim that project leaders halted the experiments due to moral concerns, fearing the misuse of timeline manipulation.

Elite Control

- Others allege that the technology is still in use but has been hidden away by elites or secret organizations to maintain control over global events.

Unfortunately (or perhaps fortunately), all these accounts are anecdotal and mostly unverifiable. Every *Looking Glass* project possesses a familiar pattern seen in other clarity crusades or truth tracking with roots in secret government operations, alien technology, or hidden knowledge—which often reflect societal fears of distrust of government, fear of the unknown, or a strong desire for humanity to achieve higher knowledge.

CHAPTER SIXTEEN

Walking The Planck (Scale) Without Wings & Time Asymmetry and Reversals

Max Planck (1858-1947) was a German theoretical physicist who revolutionized our understanding of physics by introducing quantum theory. His most famous discovery was that energy is quantized and comes in discrete packets (quanta) rather than continuous flows. His insight came from his work on black-body radiation, where he found that energy could only be emitted or absorbed in specific amounts given by E = hf, where h is Planck's constant.

In simpler terms, the Planck scale is like one of nature's "pixels"— the tiniest possible length (about a billionth of a billionth of a billionth of a billionth of a centimeter) where space itself becomes so *sploogy* that it becomes impossible to measure anything smaller.

Helpfully, the Planck scale represents a set of fundamental physical units derived from combining three basic physical constants:

- Planck's constant (h)
- The gravitational constant (G)
- The speed of light (c)

These combine to give us:

- Planck length: ~1.616×10^{-35} meters
- Planck time: ~5.391×10^{-44} seconds
- Planck mass: ~2.176×10^{-8} kilograms
- Planck energy: ~1.956×10^{9} Joules

The Planck scale is crucial to physics for several reasons including quantum gravity, because at the Planck scale, quantum effects and gravitational effects are equally important. This is where we expect quantum gravity theories like string theory to operate. Additionally, the Planck length and time represent the smallest meaningful measurements possible in our universe. Below these scales, our current physics breaks down and space-time becomes like a "quantum foam"—a roiling unknowable something that is hard to define.

Furthermore, at the Planck scale, all fundamental forces (electromagnetic, strong, weak, and gravitational) may unify into a single force—meaning, as Bohm thought—there is no separation of anything and particles as we know particles do not exist. This has implications for our understanding of electromagnetic fields and their relationship to other forces.

Current and potential applications of figuring out the Planck scale include:

A. *Quantum Computing*: Planck-scale physics helps us understand quantum coherence and decoherence, crucial for quantum computer development.

B. *Cosmology*: Understanding the earliest moments of the universe (10^{-43} seconds after the Big Bang) requires Planck-scale physics.

C. *Materials Science*: Quantum effects at small scales influence the development of new materials and nanotechnology.

A unified theory of quantum gravity able to reconcile quantum mechanics with general relativity, would offer us a better understanding of black holes and the information paradox, as Planck- scale effects become important near singularities. In addition, possible discovery of additional dimensions, as predicted by some theories that attempt to explain physics at the Planck scale, gives us new insights into the nature of time and causality at its most fundamental level.

During my research, the connection between Robert Hauschild Liebeck's *Blended Wing Body* (BWB) design principles and Planck-scale physics, highlighted an interesting convergence of aerodynamic efficiency with quantum mechanical considerations.

While the public at large is far from directly operating at the Planck scale, the BWB's integrated design philosophy aligns well with a theoretical understanding of physics at extreme Planck scales, potentially offering insights into specialized and advanced spacecraft design as scientists and engineers strive to test the fundamental limits of quantum mechanics.

Liebeck is a retired American aerospace engineer and former senior fellow at the Boeing company. He oversaw the *Blended Wing Body Program* (BWB) and is best known for his contributions to spacecraft design and the Liebeck airfoil. The concept of a blended wing body, typically associated with aviation, has intriguing implications for enhancing space or dimensional travel through its innovative aerodynamic and structural advantages.

As it relates to advanced space travel, the BWB design integrates the wings and fuselage into a seamless structure, significantly reducing drag. This improves fuel efficiency during atmospheric exit and re-entry, a critical phase of space travel.

But that's not all. The increased aerodynamic efficiency allows for more substantial payloads and smoother re-entry, making space travel more economical and reliable. The wide-body design of a BWB also allows for a larger internal space, accommodating more cargo, passengers, or scientific instruments. This is essential for long-term missions, such as establishing lunar bases or interplanetary travel.

If dimensional or interstellar travel involves passing through high-energy phenomena like wormholes or hyperspace, the structural integrity of the BWB design helps withstand these extreme forces. The integration of wings and fuselage reduces stress points that traditional designs often have.

Moreover, a blended wing incorporates advanced propulsion technologies, such as fusion-based engines or antimatter propulsion, crucial for dimensional travel, while the broad surface area of the BWB is used to integrate solar panels or other energy-harvesting systems to power long-duration missions.

On planets with atmospheres, the BWB's ability to efficiently generate lift makes it ideal for smooth landings (think Mars). This is particularly relevant for exploration missions where a controlled descent is critical (think Europa).

For spaceplanes or vehicles traveling at hypersonic speeds, the BWB design reduces heat buildup and turbulence, enhancing stability. BWB vehicles also revolutionize sub-orbital "point-to-point" travel on Earth, seriously bridging the gap and between conventional aviation and space travel. We'd likely be able to travel in a BWB from Alaska to England in twenty minutes, give or take, including boarding time, using less energy or fuel than conventional airplanes.

One of the most intriguing aspects of BWB relates to dimensional travel. At the Planck scale, additional spatial dimensions (as predicted by string theory) may become accessible. The BWB's broad, integrated design provides better stability during transitions between different dimensional states, as it evenly distributes stresses across its structure.

Moreover, the BWB's large surface area and integrated design principles would be particularly valuable for managing the enormous energies involved in Planck-scale phenomena. The design could distribute and handle the extreme energy densities (around 10^{96} kg/m^3) associated with Planck-scale events.

Relatedly, quantum computing systems within the BWB design are particularly relevant to Planck-scale physics. At these scales, quantum effects dominate. Evenly distributed quantum computational systems integrated throughout the system structure helps manage navigation and stability in regions where classical physics breaks down.

Now we arrive at human limitation, which includes scale disparity, the gap between human-scale engineering and the Planck scale. Scale disparity is so vast that any direct interaction between them requires revolutionary physics. Likewise, the energy required to probe the Planck scale (around 10^{19} GeV) is far beyond current publicly available technological capabilities—though to be sure, the BWB's efficient design principles effectively contribute to swifter incremental progress in this direction.

Way back in chapter three we examined quantum decoherence and now need to reconsider it here. Any macroscopic structure like a BWB faces significant challenges maintaining quantum states necessary for Planck-scale interactions, though its integrated design might offer definite advantages in managing decoherence.

The connection between Liebeck's BWB design principles and Planck-scale physics is an intriguing and highly coincidental convergence of aerodynamic efficiency with quantum mechanical considerations. And while the public is eons away from directly operating at the Planck scale, the BWB's integrated design aligns quite well with a theoretical understanding of physics at extreme scales.

Challenges and Research Areas

While the potential for quantum-like interstellar travel is significant, several challenges remain such as ensuring a BWB-type vehicle can withstand extreme temperatures during space travel and developing lightweight, high-strength materials to support the blended structure (which we've done).

Adapting this design to futuristic propulsion methods is a promising field of aviation and excitingly, the blended wing body paves the way for more efficient and versatile air/space travel that eventually omits wings completely.

We are on the cusp of a great human awakening. Let it be known that spheres and cube drones now fly without wings and appear to change shape with the aid of plasma technology while I continue to write fiction.

Does time exist outside of space?

We looked at spacetime in the previous chapter but is it possible for time to exist outside of space? Our best physical theories, particularly Einstein's relativity, tell us that time and space are fundamentally interconnected and form a unified "spacetime" fabric. Evidence for this includes the behavior of time dilation and length contraction, where special relativity shows that time measurements depend on relative motion through space.

In general relativity, gravity is understood as the curvature of spacetime, where mass warps both space and time together and the mathematical frameworks of modern physics treat time and space as inseparable dimensions of a single manifold. However, there are some important considerations that complicate this picture.

Quantum mechanics, as even Einstein noted, introduces some peculiar aspects where time behaves differently from space, particularly in phenomena like quantum entanglement where certain correlations appear to be instantaneous across space but still respect causality in time—taking us back to Bohm's folding and unfolding idea of the implicate and explicate order.

Subsequently, some theoretical approaches to quantum gravity suggest that at the most fundamental level, spacetime emerges from more basic entities that aren't inherently spatial or temporal and the thermodynamic arrow of time (entropy increase) seems to give time a special directional quality that space doesn't share.

While our best current understanding suggests that time and space are intrinsically linked, we can't definitively rule out the possibility that at some quantumly fundamental level they are separable or have independent existence. This remains an active area of both theoretical physics research and philosophical debate.

But what if time is not as fixed as we assume? What if time, like Leibnitz's brain research showed us in Chapter Eight, can flow backward too? Researchers at the University of Surrey, while examining how time emerges as a one-way phenomenon, discovered that opposing arrows of time do theoretically emerge from some quantum systems, where *memory kernels* are time symmetrical—and thus—time appears to move forward *or* backward.

Time Reversal Symmetry

Time reversal symmetry is another fundamental concept in physics that describes how most basic physical laws remain unchanged if time runs backwards instead of forwards.

A simple way to understand this is if you were to watch a video in reverse of a ball bouncing or planets orbiting, the events you'd see still obey physical laws. The equations that describe these processes work equally well whether time runs forward (t) or backward (-t). And this might be, as Wheeler suspected, more of a loop than a line—which means that while something like a double-slit experiment may contain super positioned particles—perhaps the choice of where a particle passes in the loop was already decided and will be decided again. And again. And again. Maybe in some worlds, time loops rather than *runs*.

There are some notable cases where time reversal symmetry breaks down like within the *Second Law of Thermodynamics*, which says that entropy tends to increase over time in isolated systems. For example, in our 3-4D world you will never see a broken egg spontaneously reassemble itself or a person age backward before your eyes.

Additionally, certain particle physics processes show a slight preference for one time direction over another (the weak force), and the measurement process appears to have a preferred time direction. This symmetry helps physicists understand conservation laws and is crucial in fields like particle physics, quantum mechanics, statistical mechanics, and condensed matter physics.

Understanding time reversal symmetry is essential in developing our modern understanding of fundamental physics and continues to be an active area of research.

Non-Linear Time

Non-linear time is a spellbinding concept that challenges our usual understanding of time as moving in a straight line from past to future. It suggests that time is sometimes experienced or structured in ways that don't follow sequential progression.

In physics and cosmology, the idea that time might curve, loop, or branch, especially in relation to Einstein's theory of relativity, shows that time is not absolute but can be affected by gravity and velocity. Other theoretical physics models, outside of Wheeler, also suggest the possibility of closed time like curves where time loops back on itself.

A way of structuring events that doesn't follow chronological order is through narrative and storytelling. Think of movies like "Memento" or "Pulp Fiction", which jump between different time periods, or novels that weave together past and present storylines through every other chapter.

Comparatively, many indigenous cultures view time as cyclical rather than linear, emphasizing recurring patterns and cycles in nature rather than progression toward a distant future. For example, Native American traditions see time as spiral-shaped, combining both circular patterns and forward movement. Could this be Wheeler's infinite loop or Bohm's *hidden variables*? Again, we face a reflection where vortexes seem to bring us around in the form of holograms and fractals.

Great thinkers like Henri Bergson distinguished between "clock time" (linear, measurable) and "lived time" (subjective, non-linear experience of duration). This relates to how we experience time where minutes feel like hours if we're bored or fearful, while hours or years seem to pass in a flash if we're having novel fun.

Likewise, meditation impacts objective experience and subjective perception of time in several fascinating ways. For example, time dilation is common during meditation sessions, where practitioners often report that time seems to pass differently (usually faster) than expected. This variability appears to be related to the depth of meditative state and type of practice being used.

The altered perception of time during meditation likely stems from changes in attention and consciousness. When we meditate, we step outside the matrix of time-bound thinking and enter a more *present-centered* state. Lab experiments have shown that the default mode network (DMN) of the brain, which is involved in self-referential thinking and time perception shows reduced activity during meditation.

Notably, regular meditation usually leads to lasting changes in how we perceive and relate to time through enhanced present-moment awareness, making it easier to fully experience the current moment rather than constantly projecting into the future or past and reduces anxiety about time passing or time pressure. It also provides a greater ability to regulate attention/emotions and choose where to direct time in relation to temporal experiences. Meditation is a more flexible relationship with time, allowing practitioners to better adapt to different temporal demands.

One particularly interesting aspect of meditation is how it helps break what researchers call "temporal discounting", a tendency to prioritize immediate rewards over long-term benefits. Through meditation, practitioners often develop a more harmonized perspective of time, helping them make decisions that better serve their long-term wellbeing. Active practitioners also tend to exhibit less suffering, loneliness, grasping, illness, resentment, or desire for instant gratification.

As we float at the threshold between classical and quantum understandings of non-linear or asymmetrical time reversals, we find ourselves confronting another quirky paradox: while the fundamental laws of physics appear largely indifferent to time's direction, our lived experience remains stubbornly linear, marching forever forward through the corridors of causality.

Perhaps it is tension itself that bars our portals to deeper insight and traversing beyond the linear arrows of time. The asymmetry we observe—from broken eggs that cannot be reassembled, to memories that accumulate rather than dissolve—may not reflect a flaw in our understanding, but rather an emergent property born of a necessary symbiosis between quantum possibility and macroscopic reality.

More likely, humans are mostly unconscious or subconscious inter-mediaries—bridges between the quantum and the classical—carrying out entanglement without awareness. Maybe consciousness is more than a byproduct of biology and is, instead, a fundamental current of the universe. As we continue to survey these human imposed bound-aries, we may find that time's true nature lies in the very question of why we must perceive boundaries at all.

CHAPTER SEVENTEEN

Warped Humor: How Gravity Tugs at the Threads of Spacetime

Spacetime Fabric

Spacetime fabric is a concept in physics that describes how space and time are interconnected as a single, four-dimensional entity. Instead of thinking about space (3 dimensions: length, width, height) and time (1 dimension) as separate things, Einstein's theory of relativity showed us they're actually woven together like a fabric. Hmmm. This sounds eerily like Bohm's interconnectedness and wholeness but there are some differences.

Oddly enough, this "fabric" can be bent (or folded and enfolded?) and warped by massive objects like stars and planets. We are shown images of this online where a heavy bowling ball is placed on a stretched rubber sheet. The sheet dips and curves around the ball. Similarly, massive objects in space curve the space-time fabric around them. This curvature is what we experience as gravity.

Some effects of space-time *fabric* occur when massive objects warp space-time and bend light rays that pass nearby (gravitational lensing). Or when time moves more slowly in stronger gravitational fields (time dilation). Additionally, massive objects like black holes allegedly create such extreme warping that even light cannot escape, and gravitational waves are ripples in the space-time fabric caused by huge cosmic events like colliding black holes.

The four dimensions of spacetime are **Length** (or width), considered the first spatial dimension. It represents movement and position along an x-axis (left/right). **Height** is the second spatial dimension, representing movement and position along a y-axis (up/down), and **Depth** is third, representing movement and position along a z-axis (forward/backward). **Time**, the fourth dimension, represents the progression of events from past to future

These four dimensions combine to form what physicists call the spacetime continuum, first formalized in Einstein's theory of special relativity. The main idea is that space and time are not separate entities but are fundamentally connected because events that seem simultaneous to one observer might occur at different times for another observer moving at a different velocity.

Think of it this way: to completely describe any event in the universe, you need to specify just where it happened (using the three spatial coordinates) and when it happened (the time coordinate). This four-dimensional framework helps explain phenomena like time dilation and the curvature of spacetime by gravity—plus it helps us get to where we are headed (on time).

Going Outside the Gravitational Model of Space-Time

In an electrical model of the universe, the idea of "bending spacetime" takes on a an entirely different meaning than it does in Einstein's general relativity. Rather than envisioning spacetime as a fabric warped by mass, this model imagines a cosmic sea of electromagnetic field lines—a dynamic lattice, which can be distorted by electric charge. When large concentrations of charge gather, they create curvatures or tensions in the field, drawing other charged particles toward them. This behavior is like how gravity pulls objects toward mass, but offers an electrical interpretation of spacetime curvature, which for me, seems to make much more sense.

So, imagine space not as an empty void, but as a fluid-like medium filled with electromagnetic energy. These fields ripple through space as waves, constantly interacting with particles. They don't just sit still—they twist like DNA into magnetic or electric forms, interchange, transfer momentum, and even reshape themselves through the presence of charge—much like how water flows and eddies around a boulder in a river over fighting against it.

Rather than mass bending space, charge seems to laser the electromagnetic structure of space, creating warps and wave-like flows that might one day help us explain gravity, inertia, or even the architecture of the cosmos itself—from the plasma-like spiral arms of galaxies to the behavior of subatomic particles.

One fascinating aspect of an electrical universe is how electromagnetic fields transform under different reference frames. What one observer sees as a purely electric field, another moving observer might see as a combination of electric and magnetic fields. This relates somewhat to Maxwell's equations, which show that electric and magnetic fields are different aspects of a single electromagnetic field tensor (a matrix, scalar, or vector having order) in spacetime.

A key difference from the gravitational model is that instead of thinking about curved spacetime *causing* gravity, we're now looking at how fields *permeate* spacetime and *mediate* forces through the exchange of photons. These electrical fields can be visualized as vectors or tensors at every point in spacetime, creating a kind of "electromagnetic filament" resonance that overlays the spacetime structure.

Electromagnetic waves are self-propagating disturbances in the electromagnetic field. When changing electric fields create changing magnetic fields and vice versa, they form a wave that can travel through a vacuum at the speed of light. Unlike mechanical waves that need a medium, electromagnetic waves *are* the medium—oscillations of the field itself. The wave carries energy and momentum through spacetime in packets called photons.

Charged particles also create electric fields that extend throughout space, diminishing with distance according to the inverse square law. When these particles move, they create magnetic fields. The fascinating part is that these fields aren't instant because changes in the field propagate at the speed of light, creating what we call *retarded* potentials. This means the field at any point represents the state of the source at an earlier time.

Interestingly, electric and magnetic fields can transform into each other through various mechanisms like moving electric charges to create magnetic fields or changing magnetic fields to induce electric fields (electromagnetic induction). Such interconversion is described by James Maxwell's equations and is the basis for most modern electronics and power generation.

At the quantum level, electromagnetic fields are quantized into photons. The interaction between charged particles occurs through the exchange of these photons when particles emit and absorb photons, fields fluctuate at quantum scales (quantum vacuum fluctuations), virtual particles briefly pop into existence, or the quantum vacuum itself has energy (zero-point energy).

This quantum view reveals that what we perceive as empty space is a seething quantum foam of virtual particles and field fluctuations or vibrations of information.

The unifying concept across all these aspects is that space isn't empty—it's filled with wave fields. These fields store and transport energy, can be warped and transformed by motion, interact with matter through charge and current, form the basis for most of the forces we experience in daily life (except gravity) and indicate quantum properties at microscopic scales.

These electromagnetic fields form a fundamental part of the structure of spacetime itself, interweaving with gravity and other forces to create the rich current of physical reality we observe and in which we actively participate.

Using Electromagnetic Fields to Manipulate Linear Time on Earth

The relationship between electromagnetic fields and time is interesting because electromagnetic fields influence our experience of linear time through time dilation effects. This happens if strong electromagnetic fields affect the local energy density of space. This means that EMF's might act as a sort of 'time brake', slowing time locally.

According to Einstein's equations, subjective human experience influences the rate at which time passes (he used the example of how we perceive time by sitting on a hot stove versus spending time with a pretty lady—where the former is too long, and the latter passes too quickly). Just as strong gravitational fields slow time, intense electromagnetic fields theoretically create similar effects, though on a much smaller scale.

Electromagnetic fields also influence the behavior of particles at the quantum level, affecting their *temporal* properties. This includes *Quantum tunneling*, where particles appear to traverse time barriers, *The Aharonov-Bohm effect*, where electromagnetic potentials affect particle behavior even in regions with no apparent field, and *Quantum entanglement*, where particles can exhibit instantaneous correlations regardless of distance.

Electromagnetic fields oscillating at specific frequencies to create standing waves in spacetime that modify local time rates, create regions of altered temporal flow, and generate interference patterns that affect how particles and energy move through time.

Theoretically, very strong rotating electromagnetic fields would create what's called frame-dragging effects, like how rotating black holes drag spacetime around them, which might result in local time distortions, closed time-like curves (theoretical paths or loops through spacetime that return to their starting point in time) and regions where time flows at different rates—EMF distortions might also give humans a sense of *missing time.*

Remember, spacetime bends under the influence of mass like the way a stretched sheet buckles under the weight of a bowling ball. The stronger the mass, the deeper the curvature, creating an effect we call *gravity.* Gravity explains why planets orbit stars and why galaxies spin in swirling formations.

Gravity does more than keep planets in motion. It stretches time, slowing it near massive objects. This strange consequence reveals that time is not an absolute force, but an undefined covert operative given a predominantly gravitational stage.

As we laser our focus into the cosmos, gravity always pulls at the fringes of our conventional understanding. From the bending of light around massive objects to the detection of gravitational waves rippling through the universe, we've blindly accepted a full extent rendering of gravity's influence over the universe, propping it up in high favor against an electrical universe. And yet, despite our sure knowledge that gravity is king, mysteries remain—dark matter, the true nature of spacetime, and what lies beyond the observable universe.

As we continue to unravel truths and mysteries, one thing is certain: gravity is a great storyteller, shaping the past, present, and future of the cosmos and physics, denying electrical superheroes their rightful place in nonfiction and a potential unified theory. And as we drift through gravity's vast reach on relativity, we remain inextricably bound by publicly available rudimentary technological advancements and fear of coloring outside of established paradigms.

PART V

Ancient Knowledge
&
Clandestine Applications

CHAPTER EIGHTEEN

The Cosmic Cubist
Metatron's Guide to Unifying Everything from Druids to DNA

R eligion and spirituality appear to seamlessly overlap when we're not paying attention. And science seems, upon first glance, to have no place in either.

However, we might compare them to mechanical and quantum physics—two frameworks that resist unification, requiring a forced alignment rather than coexisting in natural harmony. While science, religion, and spirituality aim to guide humanity, they do so with vastly different approaches and varying levels of effectiveness.

Religion (not spirituality) is the dogmatic equation set to repeat through ritual and memorization. It is sometimes monotonous and authoritatively self-interpreted as gospel. It is also replicated the world over with similar results. Like the scientific method, it has been measured, tested, and observed in what we call *the real world,* but its genuine capacity to truly save, heal, and foster genuine love are often left wanting.

Coincidentally, organized religion, like science, is typically structured and hierarchal, like an unbendable framework. It has rules, regulations, and commandments. Unlike science, it comes with a razor-wired path to an *off-world* post death paradise that requires "routes of righteousness" through daily prayers, regular church attendance, obedience, abstinence from *sin*, healthy levels of fear, and always being wary of and fighting against an enemy. In many regards, it resembles the military in its demand for uniformity, copious funding, long-term wars, and control over the fragmentation of complex systems.

On a more forgiving note, religion offers a warm sense of nostalgia, rituals, and doctrines passed down over centuries. Religion can, also like the military, whip broken souls into sobriety, or offer an opportunity to achieve effective daily function through cognitive or behavioral modifications. Religion also satisfies our sense of community and diminishes our fear of isolation. And the incense smells pretty good too.

Plus, there's Christmas, Hannukah and a host of other holidays such as Easter or Eid, where we come together with friends and family to celebrate and practice our beliefs, usually with food and gifts. On these days, we break bread and share love, and these are, for most humans, cherished centuries-old traditions.

Alternatively, religions rely on leaders—priests, imams, rabbis, chaplains, monks, and occasionally wealthy adherents—who position themselves as God's intermediaries and enforcers. These leaders ensure that God's followers interpret the "fine print" of sacrosanct terms and conditions, often (but not always) embedding subjective interpretations into moral codes for followers.

I'll admit I have issues with organized religion. It feels like a once-beloved song repeated so often that the Word has lost its sacred resonance. Group dynamics bother me too, when they inflate and shift toward controlling the masses or individuals through cataloging sins, instead of practicing nonjudgement and Love.

Unfortunately, with organized religion, it's only a matter of time before the structure—and its people—collapse under the weight of spiritual snobbery and personal suffering. I'd rather go out into the world, which is, we're told, created by God, and conduct field research, meeting people for who and where they are.

Yet admittedly, I have felt divine inspiration when I've walked into an empty church unsoiled by the negative energy of judgement, boredom, or extravagant insecurity hiding beneath luxury labels—but I've also experienced the same divine spirit on mountaintops, watching sunsets, studying flowers, sharing time with friends, or swimming in the ocean.

Then it occurred to me: My issue with organized religion lies in its separation—from spirituality, other humans, and God. Religion feels stubbornly grounded in the tangible, operating within worldly political and military-like constraints of a three-dimensional space, bound by time and materialism—much like classical mechanics and science in general.

Spirituality, on the other hand, represents elevated states of authentic and sincere connection with the divine through sensorial experience. It is both the substrate and portal to God that does not require a guru or sacred measurement meter because it is deeply personal and readily available to those who accept it through faith.

The Holy Spirit—whether understood as enlightenment, transcendence, or divine inspiration—is not drilled into the mind through programming or indoctrination. It is poured freely over the heart with love. It flings dogma to the stars and joyfully declares, *"Let's see where Source takes me!"* Having divine spirit is a vivid, open palette of wonder—perception and spacious clarity—applied with the elegance of a master artist or inventor.

Tesla, da Vinci, Oppenheimer, and Bohm were inspired by and illuminated through this sacred force, each of them having looked beyond the boundaries of science and into the realms of philosophy, music, poetry, art, and spiritual insight. In their pursuit to understand the universe—and to wield its power with humility, wisdom, and a vision for the greater good, they tried to understand the Source's seemingly conscious interconnectivity to and impact on science, themselves, and the world at large.

Spirituality is less concerned with perfection, rules, ego, arrogance, hypocrisy, or sin. If one is spiritual, he or she becomes the accountable navigator and savior, created in God's image. God becomes a co-passenger in life, helping us explore connections to the universe and what it means for humans to live well on a sentient planet.

Spiritualty, for sure, can be shared, but it, like birth and death or maybe even working in a private lab, is mostly a solo adventure—but it isn't lonely or infused with deprivation or separation—or at least it shouldn't be. Spirituality is about interconnection with others and engaging with life to successfully harness and share positive energy. It is fully realizing that if God created all beings—and is the Alpha and the Omega—then God resides everywhere, including within us—and even in the heart of our enemies—and inside the Petri dish.

And where religion feeds us acceptable answers or forces us to choose sides by insisting on a black and white form of good and evil, spirituality (and science) asks probing questions. Spirituality encourages us to love everyone, including ourselves—because in loving the self, you may love others more effectively—and honor the Creator who forged such a divine vessel in its image, through building on potentials and creating something worth sharing.

Additionally, I have rarely, if ever, seen '*love thy enemy*' in action among organized religions. It is frequently recited but rarely realized. It's one of the most difficult directives for human beings to follow: to extend compassion and love to those who harm or oppose us. Even religious institutions do not model this transcendence.

History and modern times show us that crusades, excommunications, and tribal boundaries continue to be drawn in the name of righteousness. Yet to love an enemy is the ultimate dismantling of the very architecture of ego and fear. This requires a sort of spiritual courage and humility that institutions and the people who lead them, built on hierarchy and identity, too often lack. If love is supposed to be the core of divinity, then loving your enemy, it seems to me, is the truest test of spiritual faith—and the one most ignored. If we lived by this principle instead of merely preaching it, the world would likely be a higher functioning and happier place.

Subsequently, most ancient texts, and even the Holy Bible itself, proclaims that gods (plural) made humans in the very image of their creators, meaning that God/s isn't found only when we look up—but everywhere—like a particle wave in superposition, one entangled with the human spirit and therefore, all-knowing and all-being, instantaneously.

God exists both within us and beyond us. We know God, God knows us—and in this sacred reciprocity, we come to realize that God is us, just as we are expressions of Source. It is a divine trinity of *being* comprised of God, the knower, and the known—all inseparably entwined. If we are truly God's children, as we are repeatedly told around the world, then we are not lesser fragments, but equal participants in the divine wholeness of the universe. There is no need to throw sins or science up as a roadblock to healing, spiritual growth, or forgiveness.

But here's a great thing: you don't have to pick a side. If you truly believe in God or science, your faith is everywhere. No judgment here. Remember, life's journey and your spirituality are as unique as your music playlist or field of study—and found everywhere if we walk in present awareness.

What About Free Will?

As briefly discussed much earlier, neuroscientific studies, particularly Benjamin Libet's famous experiments in the 1980s, showed that brain activity associated with movement decisions could be detected several hundred milliseconds before subjects reported making a conscious decision to move.

Later studies using fMRI have claimed to predict simple decisions up to ten seconds before any conscious awareness arose. This suggests our conscious experience of decision-making might be more a *post-hoc* rationalization (where someone creates a logical explanation for their decision *after* a choice), than actually making a choice. It might also be, as we learned in the last chapter, a minor form of subconscious quantum time reversal we don't yet realize or understand.

If we completely accept that the brain operates according to physical laws, then each neural state is caused by the previous state plus its inputs. This appears to leave zero room for "free will"—and contrasts with many religious perspectives on free will, particularly in the Abrahamic traditions.

Even quantum indeterminacy doesn't help us deal with considering there may be no free will because randomness is different from choice. This is the highly painful part for me—to consider that there is no free will—when we are cradle raised, as I was in Christianity, to fully believe in this concept and to have blind faith that if I *choose* to accept Jesus Christ as my lord and savior, it will be so.

Then it occurred to me: Maybe most humans got it wrong—and our free will—our ability to accept God into our hearts and lives, is a placebo effect based in *perception.* Before anyone despairs or arrives at my office bearing torches and wooden stakes, remember, placebos work. As we previously explored in a preceding chapter, placebos have been scientifically documented in their effects, which means that the human mind influences matter (and emotion, quantum physics, and faith).

Free will is thought of as a divine gift that sets humans apart from other creatures and is essential for moral responsibility and loving judgment. If this is true, people can be atheists and still be morally responsible, exercising good judgement. An atheist's God is within and all around them too, maybe demonstrated through science and faith in measurable experiments—and closely related to Spinoza's God.

Either way, one can refer to a higher power as the universe, say there is no higher power, or claim there is nothing but God—and each—the atheist and the believer—would be right. Quantum Physics, God, and Schrödinger's cat all bear this out.

Likewise, to describe *God* in the active, as a verb rather than a noun, Bohm's rheomode offers us a powerful lens. Instead of saying "God is love" (noun + noun), we say *"Goding is loving"*—where "Goding" implies a continuous divine unfolding or actively intelligent presence among us, inseparable from creation, perception, and being.

In this light, God is not a distant entity or static but the act of 'inter-being' —a ceaseless, generative motion. Just as Bohm envisioned reality as a *holomovement*—an undivided wholeness in flowing motion— "Goding" might be viewed as an act of universal unfolding where coherence between consciousness and matter is ongoing, the process of living an act of illuminating through a dynamic uncovering of reality. Just as we are not separate from the act of breathing, thinking, or loving, Goding is always within us, around us, and is us.

When we stop locking God into a fire-and-brimstone concept, and instead experience God as loving, creating, flowing, evolving—we loosen the grip of dogma and open to a more participatory, experiential spirituality. Bohm's rheomode thus becomes both a scientific and spiritual method for describing the *how* of God rather than the *why* of God.

Alternatively, it has been claimed that free will is independent of physical causation and necessary for the concept of sin and redemption. Is it? Some religious traditions, like certain Protestant denominations, believe in predestination. So, what now? And some scientific thinkers argue for "compatibilism"—the idea that free will is compatible with determinism if we define it carefully. Why would we have to define it carefully? Isn't the simplest path in nature and toward God usually the best? (In science we refer to this as Occam's Razor)

Critics also point out that the scientific evidence against free will isn't as definitive as sometimes portrayed because Libet-style experiments only look at simple motor decisions, not complex moral choices. Furthermore, the interpretation of "readiness potentials" in these experiments is debated, and quantum effects and chaos theory might, in fact, make perfect prediction impossible.

More importantly, *consciousness*, the enigmatic *thing* we describe but cannot define, likely plays more of a role in decision-making than these experiments capture. Even more interesting, is that some studies show that when people believe they have no free will, they behave unethically and put less effort into controlling their actions. This raises a practical question of whether *belief* in free will might be the very nature of all things and beneficial to humans, regardless of its status. Yet designing an experimental model to measure belief or free will proves practically impossible.

Ultimately, spiritualists are entangled and super positioned with this interconnection to higher states of being, free will, and power. They are free to explore, take detours, or even stop entirely to '*inter-be*'. Spirit on! Scientists, on the other hand, are bound by the rigors of the scientific method.

So, whether you're following centuries-old religious traditions to the *T*, or chasing sunsets without a compass, the real question is: what inside you, makes you feel connected, alive, and brings you heavenly peace? Because in the end, it's not about the religion or the science, it's about authenticating existence —and connecting the spiritual power of our inner light to an outlet of ethereal existence showing itself in science as anomalies—and doing so without static.

Science and Religion

Science and religion remind me of a married couple on the verge of divorce and bickering over who gets the universe and custody of "the people". Science peers through a telescope or a microscope, meticulously charting or cataloging God's creations, while religion gazes up or down with wonder and suspicion, weaving origin stories about creation and laundry listing inappropriate thoughts and behavior.

Both share a common quest: understanding the vast, mysterious "why" of existence. They just approach it with different tools—science wields data, experiments, and skepticism, while religion espouses faith, metaphor, and sin. Where science builds satellites to reach toward and explore the cosmos, religion hands down parables to make sense of the heavens.

Their differences are both their strength and their danger. Science sometimes forgets the ethical dimensions of its pursuits, rushing forward without always asking, "Should we?" Conversely, religion, clings too tightly to dogma, resisting the very change and progress that make understanding richer—and scolding anyone who might have the courage to step away from the fold to explore their faith.

Yet each have provided humanity great gifts—science has eradicated diseases, given us some of the best trauma care ever invented, and brought stars and planets closer to our understanding. On the flip side of this coin, religion inspires countless acts of astounding art, compassion, and civil disobedience, which leads to greater understanding of eternal life and connection to the divine.

Yet, unchecked, science risks becoming coldly utilitarian, and religion risks fostering division and war. Science builds the bombs that religion encourages dropped in the name of God.

But here's the beauty: when science and religion share dialogue, rather than dueling, they remind us that logic and wonder, reason and mystery, are not enemies but interconnected partners on the only planet, so far, where humans can fully and healthfully live. And perhaps their most fascinating common ground lies in the study of consciousness—the ineffable, luminous experience of "being."

What is it that makes us self-aware, capable of contemplating both the laws of quantum physics and the poetry of religious faith? What is it about the human psyche that celebrates death through the victory of war, or invites us to come together and mourn a victim's plight or build a loving community? This is where science and religion might finally fold hands together, marveling at the greatest final frontier of all: human consciousness and our interconnection to the quantum.

Spirituality & Science

Science attempts to explain the "how" of the universe, but consciousness and spirituality dare to ask the "why." Pioneers like Planck, considered the father of quantum theory, hinted at this intersection when he declared, "*I regard consciousness as fundamental.*"

In short, consciousness explores the big bang between measurable phenomena and the ineffable experience of being. It's like combining chemistry with classical music—strange but oddly satisfying. Consciousness is vast and deserves its own book. But to get to the heart of it, we should probably explore a few items and ideas where science and spirituality intersect.

Unified Field Theory: The Completely Lost Holy Grail of Physics

The Unified Field Theory (UFT) is a theoretical framework in physics that aims to describe all fundamental forces and particles of the universe within a single, unified framework. The idea is to combine gravitational, electromagnetic, strong nuclear, and weak nuclear forces—into one cohesive theory.

Physicists and mathematicians are still trying to create a single set of elegant mathematical equations that explain how all these forces interact and how the universe operates at the most fundamental level. I doubt that *thinking* our way to a unified theory will ever work. Mathematics, like consciousness, is truth—for the bubbles in which each belongs—and there exist a multitude of fragile bubbles onto which truth may be hitched.

Currently, general relativity (which describes gravity) and quantum mechanics (which describes the other three forces) are separate theories. Reconciling them absolutely is the biggest challenge in physics. A few claim to do so until proven otherwise.

Notably, Albert Einstein dedicated his life to trying to find a unified answer and warned that some parts of his theory, while fantastic, merit retooling. While we're not there yet, quantum physics and string theory tease physicists into thinking we're close. No, we are not. At least not in the public and empirical realm of science.

Certain mathematics have been classified by governments, along with certain physics. Mainstream science has been steered to focus on *acceptable* paths for unclassified research that basically and stubbornly remains inside a separate box from Schrodinger's, one that is tightly locked by gravitational models.

Yet this driven search for the ultimate pattern, understanding, cosmic depth, and perception, is the place where science meets and unifies with spirituality—every single time.

Sacred Geometry: When Math Goes Mystical

Sacred geometry refers to the belief that certain geometric patterns and shapes have symbolic, spiritual, or mystical significance. These shapes are found in nature, art, architecture, and religious symbols. The idea is that these geometries represent the fundamental laws and structure of the universe, connecting the material and the spiritual world. And many are considered archetypal, having been found in unrelated cultures all over the world.

Who knew that all those hexagons, spirals, and circles we doodled during English class turned out to be the very building blocks of existence—channeled to us through energy, frequency, and vibration in collective consciousness—the archetypal symbols coded into our very blood and DNA.

From the pyramids of Egypt to our DNA, sacred geometry shapes everything. In fact, these mesmerizing, patterns found in all matter and most art reminds us that the universe is not haphazard random chaos but structured radiance—a divine blueprint kicked into gear through resonance. If you're spiritual or even religious, it might help to think of sacred geometry as God's spirograph.

The Golden Ratio: The Universe's Secret Sauce

Nature loves harmony even within chaos, and the Golden Ratio is likely its favorite tool. Represented by the Greek letter "φ" (phi), this irrational number, approximately 1.618, appears in seashells, galaxies, and even da Vinci's Mona Lisa. It's the mathematical equivalent of symmetry—an aesthetic harmony that humans instinctively adore.

While writing this book, I took the first few numbers of phi and created a sound wave to "hear" the Golden Ratio. Each note stretched a bit beyond its ordinary and alternated between bright, warm, and muted tones—a melody born of the natural world briefly coded in infinity. This ratio tells me, intuitively, that there's an underlying intelligence guiding creation—something many of us *feel*. If the universe composed a musical score, "notes of Golden Ratio" provides its harmony.

Now, the Golden Ratio is where things really get interesting. We are told that gravity and electromagnetic forces are two entirely separate forces in nature that cannot be co-joined.

However, William Donovan, Martin Jones, and Dan Winter wrote an eloquent paper for how the golden ratio causes gravity electrically through recursive constructive wave interference—turning compression into charge as a core mechanism of self-organized and centripetal forces of gravity (*Compressions, The Hydrogen Atom, and Phase Conjugation: New Golden Mathematics of Fusion/Implosion: Restoring Centripetal Forces,* General Science Journal, 2012).

The Golden Ratio is a solution to constructive interference—and what is added may be multiplied. This completely flies in the face of damn near everything we've been taught or told regarding how our universe unfolds. How?

The idea that the universe self-organizes via a mathematical constant (like φ) appears to decentralize divine intervention from static figurehead to inter-being. Instead of a God who designs everything case-by-case, the Golden Ratio suggests the universe automatically grows and sustains itself through embedded ratios, sacred geometry, and self-reinforcing harmonics.

This implies a *God-as-conscious-system*, not *God-as-heaven-bound personality*—more akin to *Spinoza* than *Yahweh*. For the religious, this probably feels like heresy and goes against those expecting a sky dwelling-deity handing down laws and metering out thunderbolts. But don't worry, the Golden Ratio doesn't deny God—it *redefines* God as the blueprint of conscious coherence itself, coded into ratio and rhythm.

The Flower of Life

A pattern of multiple overlapping circles, which is thought to symbolize the interconnectedness of all life and the universe. Slowed in speed, when translated to musical notes, it sounds like "Om".

One fascinating and lesser-known truth about the Flower of Life is its connection to the Platonic solids, the foundational building blocks of three-dimensional geometry. These solids—tetrahedron, cube, octahedron, dodecahedron, and icosahedron—are all perfectly symmetrical and are considered the fundamental forms of matter.

But here's the intriguing part: The Metatron's Cube, a shape derived from the Flower of Life, contains within it the geometry of all five Platonic solids. These solids represent the basic forms of the physical universe but also have *symbolic* connections to the elements of fire, earth, air, water, and ether. Ancient wisdom traditions believed these forms were the "skeleton" of the universe, providing the blueprint for all energetic phenomena that created matter.

Additionally, the Flower of Life encapsulates these forms through natural emergence, where specific points are connected within its pattern. This suggests that the Flower of Life holds a geometric code for the structure of the cosmos. The pattern's universal presence— found in ancient Egypt, Peru, China, India, Tibet, Sumerian culture, and even Leonardo da Vinci's studies—hints at a universally shared human understanding of geometry, matter, and the interconnected-ness of all things.

Modern physicists sometimes draw parallels between these geometric truths and the underlying structure of spacetime, suggesting that the universe itself might resonate with the harmonies encoded in the Flower of Life, as though it isn't just simple art or math—but a conduit between the physical and quantum realms, one that humans are gifted enough to create and convey.

Fibonacci Sequence: Counting Like Nature

The Fibonacci sequence appears throughout nature and was first discovered by Indian mathematicians around 450-200 BC and named after Leonardo Fibonacci (1170-1250), first described in his book Liber Abaci (1202), as a model for rabbit population growth.

The Fibonacci sequence is what happens when numbers make art. This series—1, 1, 2, 3, 5, 8...—is a pattern where each number is the sum of the two before it. It's found in sunflower seeds, nautilus shells, pinecones, and even hurricane spirals. The Fibonacci sequence is nature's practical and poetic way of energetic resource optimization within a system.

But get this: It is linked to DNA structure too, because the dimensions of a single cycle of the DNA double helix, measuring thirty-four angstroms long and twenty-one angstroms wide, directly corresponds to two consecutive numbers in the Fibonacci sequence, resulting in a ratio that closely approximates the Golden Ratio (1.618)!

What does this mean? A single strand of DNA stores a massive amount of information. In fact, one gram of DNA can hold up to two-hundred-fifteen million gigabytes of data, making it one of the densest storage mediums on Earth. As uncomfortable as this thought might be, it will be possible, if it hasn't already happened, to create bio-computers combining both DNA with man-made circuitry. And this might be our eventual ethical undoing as humans—especially if this bio-machinery possesses even a remnant of an individual's personality but not their consciousness.

From Unity to Multiplicity

From unity comes multiplicity—everything starts as one and then splits into many, almost the way a zygote splits to become a blastocyst and then a fetus, until we have a full-blown baby human comprised of about fifty billion organized cells, where each seems to know instinctively where it belongs and what to do to realize a functioning human.

In biology, unity to multiplicity describes how a single cell divides into many specialized cells during development, while maintaining underlying interconnectedness through shared DNA, signaling pathways, and regulatory networks.

As a refresher, Bohm's framework approaches this through his concept of the "implicate order", an underlying unified reality from which the "explicate order" (our observable world of separate objects) emerges. He saw multiplicity as different manifestations of an undivided whole, like how waves are expressions of one ocean—but not separate from it. Bohm's ideas align with modern biology's understanding of organisms as complex adaptive systems with emergent properties arising from homogenous, synchronistic interactions across scales, from molecular to ecological levels.

In spiritual traditions, this concept underpins almost every creation story on the planet, where a singular *divine* source manifests into diverse forms—especially trinities such as God the Father, the Son, and the Holy Ghost in Christianity—but it also appears in Egypt with Osiris, Isis, and Horus, which represents cycles of death, rebirth, and divine kingship.

In Hinduism, Trimurti of Brahma (creator), Vishnu (preserver), Shiva (destroyer) reign while among the Celts it is said that a triple Goddess ruled consisting of Maiden, Mother, and Crone representing life cycles. In Norse mythology, Odin, Vili, and Vé are divine brothers who shaped the world.

All these trinities can be traced back to Sumerians where divine triads existed. The most notable was Anu (sky god), Enlil (air/earth god), and Enki (water/wisdom god) who together represented cosmic order. Another important triad was An, Ki (earth goddess), and Ninhursag (mother goddess). But even the Sumerian stories have their roots in much older civilizations now lost or hidden from the public.

Maybe creation stories and divine trinities are the universe's way of reminding us: we're all filaments of the same system, worshiping the same God by different names, whether we be scientists or spiritualists—or both.

Vortex Patterns: Spiraling Through Space and Time

From tornadoes to galaxies, vortex patterns are everywhere. They're efficient, channeling energy and matter in elegant spiral-like loops. In space, they appear in several forms such as accretion disks around black holes, galaxies with spiral patterns, sunspots, and coronal mass ejections.

They gather kinetic energy from the larger fluid system, like wind shear in atmospheric vortices or water currents in oceanic ones. And in some systems, such as tornadoes or hurricanes, vortices extract thermal energy from temperature gradients, converting heat into kinetic energy.

Additionally, as a vortex collects matter or energy, it strengthens itself in a positive feedback loop. For example, in whirlpools, incoming water accelerates as it spirals inward, concentrating kinetic energy.

In looking at this from a quantum perspective, in plasma physics or electromagnetic systems, vortices gather energy (this is making me think back to our earlier chapter about neutrinos) from electric and magnetic fields.

The spiraling motion aligns with field lines, concentrating electromagnetic energy into compact regions—and in some cases, vortices resonate with surrounding energy sources, such as wave patterns or oscillations, allowing them to absorb additional energy, giving rise to the idea of stable sphere-type craft being more than a possibility.

Vortices are nature's way of efficiently organizing and redistributing energy, whether in storms, galaxies, physically impossible drones, crafts or subatomic interactions.

The most exciting part is that a controlled vortex could be used to generate propellant-less propulsion by capturing its kinetic or thermal energy. Our fictional spacecraft might create a plasma vortex in a magnetic confinement field (like a tokamak used in fusion reactors). The vortex could sustain high-energy particle collisions, generating power—or—a vortex formed by ionized gases could tap into electromagnetic fields, producing thrust or powering onboard systems. When considering DNA, maybe "gifted" humans produce a vortex of spiraling cells in the body to induce Beckwith's divided space, thereby enabling greater enhancement of intelligence, consciousness, and abilities the rest of us might deem *supernatural.*

And consider a hurricane: it's basically a vortex of swirling destruction—except within its eye, where it typically remains quite calm. What if we have crafts like these, where on the outside they resemble chaotic shape-shifting, iridescent blobs but the inside is a smooth ride?

The Vesica Piscis

Formed by the intersection of two circles, this shape is often seen as a symbol of duality, creation, and the union of opposites. What I find so interesting about this simple geometry is its potential for wave interference and superposition. For instance, the overlapping circles might symbolize quantum superposition and wave interference.

Remember, in quantum physics, particles exist as probability waves that overlap, much like the Vesica Piscis shows a region of intersection. This could serve as a visual metaphor or even a model for analyzing wave functions in quantum mechanics.

Furthermore, the Vesica Piscis represents *connectivity* between two systems. It might be used to describe entangled quantum states, where two particles are interconnected regardless of distance, analogous to the overlapping region binding the two circles.

Interestingly, this figure embodies proportions foundational to geometry and mathematics, such as the square root of two and the Golden Ratio. These constants could be key in describing spacetime geometry at quantum scales, where classical Euclidean geometry fails.

Going even further, this beautiful symbol might inspire designs for quantum gates (the basic units of quantum computers). Its symmetry and inherent proportions might also assist in optimizing how quantum states are manipulated. Yet the most interesting part for me regarding this symbol is its potential for energy resonance and amplification, spacetime manipulation, light manipulation and advanced communications with otherworldly or interdimensional beings.

What if advanced extraterrestrial technologies or even humans harnessed this geometry for devices that amplify or resonate energy fields, possibly for propulsion or communication? Now, it's true, I'm getting ahead of myself in the Sci-Fi department—but why not? A few of our governments have already beat us to it, by decades in some cases.

Moreover, if advanced extraterrestrial civilizations somehow mastered spacetime engineering, could they use the Vesica Piscis geometry to create or manipulate existing wormholes or warp fields? *Could we? Do we?*

Vesica Piscis geometry is highly relevant in optics, where light interference patterns are critical. Advanced quantum computing or extraterrestrial technologies might employ similar geometrical configurations to harness and guide light for communications, energy systems, surveillance, or even invisibility cloaking—or maybe it's already being used as a communication bridge—perhaps a functional element in some sort of universal cosmic or interdimensional language.

Ultimately, what is so appealing about the Vesica Piscis is its ability to connect mathematical abstraction with practical application. This makes it a fascinating candidate for quantum and extraterrestrial technological exploration.

Celtic Druids' Twelve Circles: Mysticism Meets Math

The Celtic Druids did a lot more than hug trees. The problem is—they left little behind in way of written materials or tools from which we might learn more about them. We can speculate with a degree of semi-certainty that they were geometry buffs. Their twelve-circle system is said to encode sacred knowledge about the cosmos, possibly mirroring lunar cycles, seasons, and even the zodiac in places like Stonehenge or other sacred sites.

The Celtic concept of the twelve circles is sometimes linked to sacred geometry but more with mystical cosmology. Yet it offers intriguing parallels with ideas in quantum physics—especially Bohm's idea of interconnectedness.

Twelve circles symbolize interconnectedness, much like quantum entanglement, where particles remain linked regardless of distance. Maybe the Druids saw their circles built all over Europe as an integrative map of the universe, reflecting the quantum principle that no part of a system exists in isolation. Each circle might represent an "aspect of being" in a cosmic matrix of relationships, like entangled states.

The important thing to remember about any sacred geometry is its emphasis on symmetry, which plays a crucial role in quantum mechanics.

The Druids' twelve circles might mirror the idea of higher dimensional symmetry, a key to understanding phenomena like particle behavior, which operates in multidimensional spaces.

Similarly, the Druids seemed to use their circles to represent unity and the multiplicity of existence. Quantum physics echoes this with wave-particle duality, where light and matter exhibit both unified (wave-like) and distinct (particle-like) properties. The circles might align with how the universe transitions from the indescribable, non-measurable substrate of a unified quantum field to the multiplicity of observable phenomena.

Additionally, the cyclical nature of the twelve circles resonates, for me anyway, with quantum behaviors like oscillations or the periodicity found in atomic structures. Metaphorically speaking, these cycles and circles might have served as links to help the Druids better understand time, space, and energy flows.

Interestingly, the geometric arrangement of twelve circles usually incorporates the Golden Ration and harmonics, which are observed in both natural patterns and quantum phenomena. Systems achieving coherence (e.g., in lasers or Bose-Einstein condensates) exhibit similar harmonics—and—the Druid circles appear to reflect this coherence, emphasizing alignment and harmony in the universe. As the Druidic Circles go, so too, may the crop circles popping up all over Europe. Maybe the Druids successfully contacted something or someone centuries ago—and we are the current recipients of interdimensional or extraterrestrial (or even under-terrestrial) messages—or our government and its allies are practicing their extraordinarily advanced laser skills on farmer fields, while simultaneously conducting successful psyops experiments.

In some spaces, the Druids were suspected of studying telluric currents—natural flows of Earth's energy like quantum fields permeating space. Each, the physicist and the Druid examined forces that influence matter and energy, albeit in highly different ways. Maybe the Druids were onto the multi-verse theory centuries before modern physicists.

While the Druids' mystical interpretation of the twelve circles was likely symbolic, it offers an alternative lens for imagining quantum physics and sci-fi. Each approach hopes to decode the fundamental structure of reality, one through spiritual geometry, one through mathematics and experimentation and the other through imaginative story telling. Bridging these ideas invites a much more interesting, playful, and interdisciplinary understanding of the universe.

Caduceus Symbol: Snakes and Ladders of Healing

Two snakes coiling around a staff (sometimes topped with wings)—the caduceus is the universal symbol of healing and medicine. Its origins trace back to Hermes, the Greek god of transitions. Spiritually, it represents duality (masculine and feminine, yin and yang) and the human spine's kundalini energy—in other words, it symbolizes duality and balance, such as life and death, chaos and order, or illness and health—which ultimately—is all about harmony.

If everything remains the same and there is never any change, nothing happens—or maybe everything that did happen eventually withers. If we have no rain, there is no food. If we have no sun, there is no life. If we have too much sun, everything withers. Even hurricanes or wildfires, as painful as they are in bringing loss and suffering upon the landscape and its people, awakens humans to glorious acts of compassion and moments of reflection that involve improvement and change—nature's often unexpected and seemingly unfair way of pruning.

I'd seen the Caduceus symbol as a child, giving it little meaning except when I had to visit a doctor. The symbol entered my life again only recently, when I came across the "lost zodiac" sign of Ophiuchus.

While Ophiuchus is not traditionally included in the twelve-sign zodiac system, it has gained attention as a 13th discarded astrological sign, that disrupts our familiar astrological calendar.

It was the Babylonians who chose to exclude Ophiuchus for symmetry and convenience in measuring time, which made their system more straightforward for use in agriculture, navigation, and religious rituals. We stuck with their twelve zodiac symbols into modern times.

Ophiuchus is associated with the Greek healer Asclepius, who, it was said, could bring the dead back to life. Asclepius learned the secrets of healing by observing serpents, which were once viewed in polytheistic traditions as symbols of wisdom, renewal, and transformation.

The Christian church sought to stamp out polytheism by having snakes rebranded as the devil's favorite evil shapeshifting form—and ordered them killed by the thousands. I can't find any evidence of it, but I've always wondered if the mass killing of snakes enabled rats, and later, Bubonic plague, to thrive.

The constellation Ophiuchus depicts a man holding a serpent, dividing the serpent's body into two parts. This imagery connects Ophiuchus to themes of flowing mastery into enlightenment over primal forces, duality, and the pursuit of higher knowledge—it encourages curiosity and asking questions, understanding our interconnectedness, and revering the great Source from which all spirituality springs.

It is also a legitimate constellation located near the celestial equator. The Sun passes through Ophiuchus from November 29 to December 17, making it a contender for inclusion in the zodiac calendar. Those born under this hidden sign are considered empaths, healers, and seekers, according to astrologers who even know about Ophiuchus. They are said to be transformational types, inspiring change in others. Ophiuchus' are considered quantum alchemists who highly value their independence—and were thrown under the sign of Sagittarius to make things easier from a calendar perspective.

From a physics perspective, I see two snakes entwined around the staff as a symbol of entangled particles — seemingly separate entities bound by invisible, non-local connections.

The Caduceus's shape resembles a waveform too, tying it to the idea of healing through vibrational energy or resonant frequencies, which are bona fide scientific things. For me, the two snakes represent the dual states of quantum bits (qubits), embodying both possibilities simultaneously in superposition.

Since I like to play around with science fiction ideas, the lost sign of Ophiuchus might represent a gateway to understanding hidden forces or energies of the cosmos, akin to unlocking dormant DNA sequences or harnessing dark energy fields.

This convergence of the Caduceus and Ophiuchus also opens imaginative portals in my mind where ancient symbols transcend their origins to represent concepts of unity, balance, and the hidden mechanics of the universe.

Metatron's Cube: The 3D Blueprint of Creation

Think of Metatron's Cube as the universe's Rubik's Cube—solving it reveals cosmic secrets. It is the sacred geometry all-star, a 3D symbol born from thirteen circles and their interconnections, derived from the Flower of Life. This intricate pattern is said to contain all Platonic solids (tetrahedron, cube, octahedron, dodecahedron, and icosahedron), which are considered the building blocks of three-dimensional space.

Named after the archangel Metatron, the cube represents harmony and the interconnectedness of all things. Metatron is tasked as a celestial scribe who records the choices made by divine and earthly beings in the Book of Life. He also serves as a guide to humanity, an intermediary between God and humans, known in Christian, Jewish, and Islamic religions. His name was once Enoch—and he used to be human before being taken to the heavens by the Gods (plural). His name in Sumerian (appearing about 4,000 years before Christ), as a first priest, was Enmeduranki from Sippar.

The patterns in Metatron's Cube reflect the geometric fabric of space-time, like the way quantum physics describes particles and fields in terms of waves and symmetries. The interconnectedness of its lines and circles symbolize entanglement or the relationship between particles across space.

Similarly, in quantum physics, everything vibrates at certain frequencies. Metatron's Cube can be interpreted as a visual representation of harmonic resonance, with its shapes reflecting the stable forms matter can take under specific vibrational states, the way sand or water does on plates when exposed to certain vibrational frequencies (or even words). If the vibrations are positive, gorgeous patterns appear. If the words or intentions are negative or unpleasant, water and sand patterns appear to lack symmetry, or experience what I term *troubled coherence*—and looks disheveled, as if it's been abused.

Going further into speculation, maybe Metatron's Cube metaphorically aligns with quantum fields, where the lines represent the energetic pathways or force connections between particles or quantum states. Its ability to contain and unify different forms might also be linked to the concept of multiple dimensions in quantum mechanics.

Science fictionally speaking, Metatron's Cube might serve as a "map" for interdimensional travel. Its precise geometry could encode the coordinates for traversing wormholes or accessing higher dimensions—or—the figure could be imagined as a blueprint for advanced energy systems, harnessing zero-point energy or creating stable energy fields for spacecraft. What if a spaceship drive could use Metatron's Cube as the core for a quantum reactor, allowing near-infinite energy by aligning with cosmic vibrations?

And what about conscious mapping? The cube could symbolize the architecture of a hyper-advanced biotechnical AI, where each connection within it represents a *neural* pathway in a quantum computer. Likewise, for war and defense, the cube's geometry could inspire designs for energy weapons or shields, leveraging its Platonic Solids to stabilize or amplify energy fields.

Or maybe, like *Project Looking Glass*, the cube could serve as an interface to a higher dimension, enabling communication with non-corporeal entities—or space-time hopping points where we could see into and possibly influence the present, past or future.

Metatron's Cube provides fertile ground for exploring the intersection of sacred geometry, quantum theory, and imaginative science fiction, offering a potent speculative and highly interesting framework for advanced technology.

CHAPTER NINETEEN

Ancient Wisdom Traditions (AWT's)
Old Is Gold

W hether it's Vedic or Jesuit monk chants or Native American rituals, ancient wisdom traditions remind us that the past and our history holds important information for our present and future. Sacred and cosmic knowledge exists within a host of traditions, and each understood the interconnectedness of all things long before science did.

While sometimes cryptic or secretive, ancient wisdom traditions are like the great-great-grandparents of philosophy, religion, and technology, and worth tuning into to gain greater experience, knowledge, scientific advancement, and for enhancing creativity and intuition.

Ancient wisdom traditions (AWT's) are typically philosophical, spiritual, and cultural systems developed by civilizations throughout history to understand life, the cosmos, and humanity's place within them. These traditions integrate cosmology, ethics, metaphysics, and practical teachings on living and communicating harmoniously and authentically with oneself, others, Earth, and the universe.

Like Bohm's *Hidden Variables,* AWT's emphasize interconnect-edness—among humans, nature, and the cosmos—encouraging an integrative and holistic worldview that contrasts with modern society's fast-paced, fragmented, highly reductionist approaches. Researching AWT's inspire solutions to contemporary problems like overcrowding, drought, flooding, shelter, food, or even social disconnection.

Similarly, practices such as meditation, mindfulness, and yoga, rooted in ancient traditions, have been scientifically validated to reduce stress, improve mental health, connectedness, and enhance quality of life—a counter-balance to the fast-paced, high-stress nature of modern living. AWT's offer tools for personal resilience and emotional harmony.

Notably, indigenous traditions teach respect for the Earth and sustainable ways of living, which are crucial as humanity grapples with the fallout of magnetic pole shifts, increases in severe weather, and rising sea levels—partially our fault but mostly natural order. Reviving these ancient principles might serve as a guide to modern ecological practices and policies.

Furthermore, ancient teachings, like those in Buddhism, Confucianism, Stoicism, or Indigenous wisdom, provide humanitarian guidelines, which might address moral dilemmas in areas like technology, artificial intelligence, and global governance. For example, philosophies such as those in the Bhagavad Gita, The Art of War by Sun Tzu, Meditations by Marcus Aurelius, Rumi, or the Tao Te Ching offer timeless insights into ethical leadership and decision-making.

Subsequently, ancient wisdom integrates non-linear, intuitive, and experiential approaches to knowledge, which complement modern science and innovation. Fields like quantum physics, systems thinking, and holistic medicine increasingly echo principles found in ancient traditions, such as interconnectedness and non-duality. By engaging with these traditions, we honor the richness of human history while finding shared principles that unite us.

From a scientific perspective, ancient geometrical knowledge, cosmologies, and metaphors inspire advancements in fields like architecture, design, biomimicry, and even quantum physics.

In essence, ancient wisdom traditions offer a link to the past and a newly designed compass for navigating the complexities of modern life. By integrating AWT's with contemporary knowledge, we build a more resonant and meaningful future.

Hinduism (Vedas and Upanishads)

- Origin: India
- Focus: Cosmic order (Dharma), the interconnectedness of all life, and the pursuit of enlightenment through meditation, yoga, and devotion.
- Key Ideas: Karma (actions have consequences), Moksha (liberation of suffering and union with the divine), and the cycles of Samsara (rebirth or suffering).
- One of my favorite things about Hinduism is the idea that spirit resides in everything—even the mundane, where even tasks Westerners might deem tedious or difficult are typically met with a sense of stoic duty—a form of worship of the Source for what makes matter manifest.

Taoism

- Origin: China
- Focus: Living in harmony with the Tao (the Way), emphasizing balance, simplicity, and flow (harmony) with nature.

- Key Text: *Tao Te Ching* by Laozi.
- Practices: Qi Gong, Tai Chi, and Feng Shui.

Buddhism

- Origin: India/Nepal/Tibet
- Focus: Overcoming suffering through mindfulness, ethical living, and meditation to attain enlightenment.
- Key Ideas: The Four Noble Truths (the truth of suffering and its cause, end, and path) and the Eightfold Path (rightful view, action, speech, resolve, livelihood, effort, mindfulness, and concentration).

Native American Wisdom

- Origin: Various indigenous tribes across the Americas
- Focus: Deep respect for nature, the cycles of life, and communal living.
- Key Ideas: Animism (consciousness exists within everything), balance with the Earth, and the wisdom of ancestors and spirits.

Kabbalah (Jewish Mysticism)

- Origin: Israel/Mediterranean
- Focus: Exploring the mysteries of God, creation, and the soul—understanding God helps us understand ourselves
- Key Texts: *Zohar* and the Tree of Life.
- Practices: Contemplative study, meditation, and symbolic interpretation, particularly as it pertains to harmonics

Bon Tradition

- Origin: Tibet (pre-Buddhist)
- Focus: Connection with the natural world and the unseen realms, blending shamanic practices and spiritual teachings.

- Practices: Rituals to align with cosmic energies and protect against negative forces.

Sami Shamanism

- Origin: Northern Europe (Scandinavia)
- Focus: Connection with the spirit world through trance and nature rituals.
- Practices: Use of drums and chanting to communicate with ancestral spirits and nature deities.

Ifá Divination (Yoruba)

- Origin: West Africa (Nigeria, Benin)
- Focus: Understanding destiny and aligning with cosmic forces through Orisha (divine beings).
- Practices: Divination, dance, and music.
- Influence: Strong ties to Santería and Candomblé in the Americas.

Ainu Animism

- Origin: Japan (Hokkaido region)
- Focus: Reverence for Kamuy (spirits) inhabiting nature.
- Practices: Ritual offerings to ensure harmony and mutual respect with the spirit world.

Mazatec Shamanism

- Origin: Oaxaca, Mexico
- Focus: Use of sacred plants (like psilocybin mushrooms) for healing, insight, and connection with divine wisdom.
- Practices: Guided ceremonies with chants and prayer.

Druidry

- Origin: Celtic Europe
- Focus: The interconnectedness of life, reverence for nature, and the cycles of the seasons.
- Practices: Rituals honoring solstices and equinoxes, storytelling, and tree worship.

Zoroastrianism

- Origin: Persia (modern-day Iran)
- Focus: Duality of good and evil, with an emphasis on truth, order (*asha*), and the sacredness of fire (divine wisdom)
- Key Texts: *Avesta*
- Disposal of dead through exposure and nature rather than burial

Aboriginal Dreamtime

- Origin: Australia (it is over 65k years old)
- Focus: Understanding the spiritual origins of the world and the interconnection between land, beings, and ancestors.
- There exists a spiritual dimension parallel to the physical world, accessible through ceremony, dance, art and ritual.
- Practices: Songlines, storytelling, and sacred art.

Huna (Hawaiian Wisdom)

- Origin: Polynesia (Hawaii)
- Focus: Alignment of the body, mind, and spirit with divine energy (*mana*).
- Practices: Ho'oponopono (healing through reconciliation) and Kahuna (spiritual wisdom keepers). Kahunas might be master

carvers or prayer warriors or have healing gifts. They might also be forecasters or star readers.

Hermeticism

- Origin: Greco-Egyptian (based on texts attributed to Hermes Trismegistus, *Hermetica (corpus hermeticum)*, a syncretic fusion of the Greek god Hermes and Egyptian god Thoth)
- Focus: Universal truths expressed through the principles of correspondence, vibration, and polarity
- Gave rise to spiritual alchemy
- Influence: Alchemy and Western Hellenistic and Egyptian esoteric traditions
- "As above, so below"

Interestingly, Hermeticism's focus on polarity and vibration sounds a lot like something we'd find in quantum physics:

- Everything is in constant motion/vibration.
- Different rates of vibration create different manifestations of matter and energy.
- Even seemingly solid objects vibrate at an atomic level.
- Higher vibrations = spiritual/mental states; Lower vibrations = physical matter.

Polarity:

- Everything has an opposite aspect
- Opposites are identical in nature but different in degree (hot-cold, light-dark, love-hate)
- Understanding this polarity of opposites enables transformation between the states
- Humans can transcend duality through understanding unity

The *Kybalion,* published in 1908, is said to be based on *Hermetica,* but appears to be more influenced by 19th century New Thought than the ancient text. However, it does mention polarity, "as above, so below", includes karmic references, and claims that the universe is made up of mind (consciousness) that gives rise to all matter.

Telluric Currents: Earth's Energy Grid

These natural electric currents flow through the ground or underwater, like Earth's nervous system, influencing everything from geology to human health. Ancient cultures tapped into them to align sacred sites, like Stonehenge, Machu Picchu, and the Great Pyramids. It's proof that the Earth isn't just an unfeeling, submissive rock; it's alive and buzzing—and humans and animals can connect and interact with this energy.

Variations in Earth's magnetic field, conductivity differences in the ground, and interactions with solar and cosmic radiation influence telluric currents. TC's are typically low-frequency and closely associated with geomagnetic activity, including auroras and geomagnetic storms. Druids and other shamans around the globe allegedly conducted rituals to harness *Earth energy,* tapping into these subtle electromagnetic fields.

Additionally, Chinese Feng Shui incorporates ideas that align with telluric currents, emphasizing how Earth's energy and its flow influences well-being, harmony, and prosperity. In Europe and the Middle East geomancy still uses a form of divination techniques to identify favorable locations for construction, oil reserves, natural gas, mineral deposits, and water, influenced by natural currents.

Interestingly, many cultures symbolized the interaction of Earth's energy with the heavens using sacred geometry, such as spirals or patterns like the Vesica Piscis, which align with energy flow concepts.

As it pertains to quantum physics, telluric currents might theoretically exist on other planets with magnetic fields and conductive materials, providing insights into planetary geology and subsurface processes. Closer to home, telluric currents serve as natural proxies for studying the planet's interior, aiding research into planetary magnetism. And at a quantum level, these currents interact with Earth's geomagnetic field, offering a natural laboratory to study electromagnetic phenomena in real-time.

Subsequently, quantum coherence or entanglement concepts might be investigated in systems influenced by Earth's electromagnetic environment, especially in particle and wave propagation studies. It's possible too, that these currents and their behavior, could provide impact data of solar activity on Earth's magnetic field, which is crucial for satellite technology, space navigation, and long-distance communications.

Nikola Tesla somehow knew that Earth's conductive properties allowed the wireless transmission of power—an idea with conceptual roots in telluric currents. He said, repeatedly, that many of these ideas came to him as *visions*. This idea has inspired other research into low-energy quantum systems.

Low-energy quantum system research spans numerous fields, with applications in quantum computing, condensed matter physics, quantum optics, and fundamental physics. These studies aim to explore quantum phenomena at energy scales where thermal and classical effects are minimized, revealing subtle quantum behaviors. Superfluidity, quantum biology, superconductivity, and insights into zero-point energy are but a few that come to mind.

Likewise, some quantum physics frameworks propose that Earth's electromagnetic currents create resonance patterns that affect consciousness or even quantum states at the macro level. While still controversial, these ideas overlap with ancient wisdom that links energy fields to human awareness, and according to Bohm's *Hidden Variables*, may not be as far-fetched as once thought.

Yet again, the integration of ancient concepts with modern scientific understanding of telluric currents opens pathways for sustainable energy research using Earth's natural electric fields, space exploration technologies that mimic Earth's natural electromagnetic dynamics, and quantum research inspired by the harmonious balance seen in ancient geomantic practices.

The interplay between telluric currents, geomagnetic fields, and the cosmos continues to inspire both scientific innovation and a deeper appreciation of Earth's naturally dynamic systems.

Universal Symbols: The Alphabet of Existence

From the yin-yang to the Tree of Life, universal symbols are humanity's way of capturing the ineffable. Across cultures and millennia, these icons speak a common language, hinting at shared truths—and there are many. Like emojis for the soul—universal symbols abound, are considered powerful, and are instantly recognizable because they co-mingle human experience with the natural and cosmic world.

I've listed only a scant few of many archetypal or symbolic representations that recur across cultures and time:

Nature & Elements

1. **Sun** - Life, energy, creation
2. **Moon** - Intuition, change, cycles
3. **Star** - Guidance, aspirations, the cosmos
4. **Earth** - Fertility, stability, home

5. **Water** - Flow, cleansing, life force
6. **Fire** - Transformation, energy, destruction
7. **Air/Wind** - Freedom, intellect, spirit
8. **Tree** - Growth, connection, life
9. **Mountain** - Strength, stability, spirituality
10. **Ocean** - Depth, mystery, vastness

Geometric Symbols

11. **Circle** - Unity, wholeness, infinity
12. **Triangle** - Balance, change, direction
13. **Square** - Stability, order, foundation
14. **Spiral** - Evolution, growth, journey
15. **Hexagon** - Harmony, interconnectedness, natural order
16. **Vesica Piscis** - Creation, duality, balance
17. **Flower of Life** - Sacred geometry, life force, interconnectedness
18. **Labyrinth** - Path, journey, self-discovery

Archetypal Symbols

19. **Eye** - Awareness, insight, protection
20. **Heart** - Love, emotion, life force
21. **Cross** - Intersection of planes, spirituality, sacrifice
22. **Caduceus** - Healing, balance, duality
23. **Ankh** - Eternal life, spirituality
24. **Om** - Sound of the universe, consciousness
25. **Lotus** - Purity, enlightenment, spiritual awakening
26. **Yin-Yang** - Balance, duality, harmony

A Limited Number of Animal Symbols

27. **Snake** - Transformation, healing, cycles
28. **Eagle** - Vision, strength, freedom

29. Lion - Courage, power, leadership

30. Dragon - Power, protection, wisdom

31. Butterfly - Transformation, rebirth, beauty

32. Dove - Peace, love, hope

33. Wolf – Healing, protection, guardians, guides

34. Lamb—Innocence, purity, sacrifice

No discussion of symbols would be complete without mentioning the infinity symbol (∞), also known as the lemniscate. Rich in meaning, it has been used across different fields—from mathematics to spirituality, and even in quantum physics, to signify *forever*.

In the context of time and space, the infinity symbol is used to denote cycles of eternity, which suggest time isn't linear (past → present → future) but looping. Einstein's Theory of Relativity also agrees that time is not absolute but intertwined with space. The curvature of spacetime around massive objects creates phenomena like time dilation, where time stretches or contracts depending on the observer's speed or gravity's influence.

Spiritually, the infinity symbol is a representation of universal interconnection, including balance, harmony, karma, reincarnation and the universe's renewing nature. Quantum physics brings some compelling, though complex, ideas to the table relating to the infinity symbol through entanglement, superposition, and multi-verse theory.

The infinity symbol integrates science, philosophy and spirituality. Whether it's representing the unending fabric of spacetime, the soul's eternal journey, or the infinite possibilities within quantum states, it holds a universal truth right at the point where boundaries between dualities such as life and death, known and unknown, and spacetime, blur.

Ultimately, ancient wisdom traditions inform us about our history and help us understand the unseen forces shaping our existence. Across cultures, symbols like the spiral, the tree of life, and the ouroboros endure—not as arbitrary images but as signs pointing toward quantum truths.

Consider telluric currents—natural electric currents flowing through the Earth's crust. Indigenous shamans, Vedic sages, and Taoist masters all spoke of and utilized energy lines, ley lines, and vibrational pathways. They mapped energy fields in a way science is only recently and reluctantly beginning to measure. Ancient mystics allegedly tapped into telluric forces to navigate consciousness, space, and time—and many temples and churches were, centuries later, built over such sites. What might we rediscover by tuning into these ancient teachings?

Wisdom, I'm learning, is not just stored in books or traditions, but in the very Earth beneath us, the nature around us, and the among the skies above us. See. Listen. Investigate. Dream. Could the answers to our existence, past, and future be found in decoding our history by imagining it as a living, breathing system of interconnected wisdom?

As modern civilization rushes toward an uncertain and high-risk future, we should ask ourselves: Can we afford to ignore the signals encoded in ancient traditions and the very currents and shifting patterns propelled above, around, and beneath our feet?

CHAPTER TWENTY

Mystery Schools &
Secret Societies

Entire books, wonderful works, have been written about mystery schools and secret societies, which have existed throughout history. Each has primarily served as repositories of specialized knowledge or centers of spiritual development for the elite classes in largely 'first-world' *advanced* societies whether contemporary or ancient.

In ancient Egypt, the Mystery School of Isis was one of best known, teaching initiates about consciousness, natural law, and divine wisdom. The teachings were carefully guarded and revealed gradually through levels of initiation. The cyclical nature of life, death, and rebirth emphasized the power of *feminine* energy within this school. The importance of personal transformation through initiation rituals, and the potential for spiritual healing, and connection to the divine through the figure of Isis, represented the nurturing and protective aspects of the goddess archetype.

In ancient Greece, the Eleusinian Mysteries were particularly significant, operating for nearly two millennia. These initiation ceremonies honored Demeter (agriculture and fertility) and Persephone (changing seasons), focusing on the cycle of death and rebirth. Initiates were sworn to secrecy, though we know they experienced powerful ritualistic ceremonies that led to substantial psychological and spiritual transformations. The goddesses, considered mother and daughter in mythology, were typically worshiped together to represent birth, life, death and rebirth.

The Pythagorean school, open to both men and women, while known for mathematics, was a mystery school. It viewed numbers as key to understanding reality. Its initiates combined mathematical study with spiritual practice, believing that reality's deepest secrets could be understood through numerical relationships. The Pythagoreans, for example, displayed an interest in metaphysics, though they claimed to find its key in mathematical form rather than in any substance.

The rules for the religious life that Pythagoras, its founder, taught were largely ritualistic: refrain from speaking about the holy, wear white clothes, observe sexual purity, and curiously, do not touch beans. He and his followers believed that purification could be found by elevating oneself closer to the divine creator through music and philosophy.

The Knights Templar were officially known as the *Poor Fellow-Soldiers of Christ and of the Temple of Solomon*, founded around 1119 CE during the Crusades. Initially established to protect Christian pilgrims traveling to Jerusalem, they evolved into a powerful communistic sort of military and financial organization. The Templars essentially created one of the first banking systems, where pilgrims deposited money in a European Templar house and could withdraw it in Jerusalem using a coded letter of credit.

Its members underwent rigorous initiation and followed a strict code of conduct, including sharing everything and living a communal existence. They abided by something called the *Latin Rule,* seventy-two clauses of existence, which restricted meat, imposed silence during meals, mandated austerity, zero contact with women, including hugs from family members, and the wearing of white cloaks.

The dramatic end of the Templars came in 1307 when King Phillip IV of France, deeply financially indebted to the Order, had its members arrested for heresy and its leaders burned at the stake.

The ***Order of Skull and Bones***, founded in 1832 at Yale University, is perhaps the most famous. Known as "The Order" or "322," it has included influential members including multiple U.S. Presidents. Annually, only fifteen new members are inducted, and meetings allegedly occur in a building known as *The Tomb.* Each male only member receives a nickname based on a historical figure and is assigned a societal role based on that figure.

Death imagery is extensive throughout the Skull and Bones order and in its symbolism, such as allegedly lowering initiates into a pit of snakes or locking them in a coffin for three days where they are eventually released "born again". It is considered a psychological experience to build loyalty to the group (even above spouses and children) and to maintain secrecy. Some of its members believe they are chosen and anointed by God to rule the world.

The Death's Head (Totenkopf) was another skull-based society, used by various Prussian military units and later adopted by other organizations. The symbol represented the motto of *fearlessness in the face of death.*

More than one-hundred years later, during the early days of the Nazi Party, Julius Schreck, leader of Hitler's personal security, resurrected the use of the *Totenkopf* as the unit's insignia, which later grew into the SS. According to writings by Heinrich Himmler, the Totenkopf served as a reminder that followers would always be willing to take a personal hit (even death) for the whole, to ensure its continuation. Alternatively, the German Nazi Swastika was different and represented nationalism and the racial superiority of *Aryans*.

However, the swastika existed hundreds of years before it was misbranded in Germany. The word "swastika" comes from the Sanskrit word *svastika*, which means "good fortune" or "well-being". It is still used by Hindus and Buddhists today as an implement to remove obstacles or as the embodiment of prosperity and good fortune. In Hinduism, it is associated with Lord Ganesha, a preserver of esoteric knowledge.

The '*Wahrheitsgesellschaft*' (Society for Truth) was founded in Pre-WWII Berlin and devoted its time searching for Vril, or *secret energy*. Most of its beliefs were likely based on a fiction novel, *The Coming Race* published in 1871 by the English writer Edward Bulwer-Lytton. His book highlights a secret antediluvian subterranean society superior to humans who possess vital psychic energy.

German rocket pioneer William Ley published a 1947 article in *Astounding Science Fiction* where he claimed that Bulwer-Lytton's subterranean beings were a hoax...but the Vril was not...Vril came in the form of small metal spheres known as *Lares* (Lares were guardian deities in Roman religion whose origins are unknown). Not surprisingly, Vril was pushed to the societal fringes and dismissed as either silly fiction, nonexistent, or propaganda —while the Thule Society was a post WWI occult and ethnic "blood and soil" nationalist group out of Munich, based on a mythical Greek country.

Curiously, I've always wondered if the *Lares* were man-made spheres the Allies referred to as *foo fighters* during WWII.

Thule was reorganized by Adolf Hitler into the National Socialist German Workers Party (Nazi Party). The organization's membership includes plenty of Nazi sympathizers like Rudolph Hess, Alfred Rosenberg, Gottfried Feder, Hans Frank, Julius Lehmann and others.

The Thule Society was initially touted as a "German study group" headed by Walter Nauhaus, an occultist who belonged to the *Germanorden* or Order of the Teutons, founded in 1911. The hierarchal structure was based on freemasonry with the idea that Jews engaged in secret organizations so they should too.

Some Thule members were convinced that Islamic and Germanic mystical systems shared a common Aryan root. (*Aryan* is a concept that arose in the late-19th century to describe people who descended from Proto-Indo-Europeans. The terminology historically means *Iranian noble*).

All of this leads me to the Nazi bell-shaped UFO, otherwise known as Die Glock. Everyone from truth crusaders to government whistleblowers claim that the device was for time-travel, harnessing energy, or that it was some sort of mega weapon. It existed. It was unstable. And it was *DEEPLY* classified by the American and UK governments.

One fascinating connection between all these organizations is their use of death symbolism as a tool for transformation. The skull and crossbones represent both mortality and the idea of transformation—a relinquishing of one's old self to be reborn into new unbreakable loyalty, knowledge, and greater esoteric understanding. Symbolic death or being 'born again' is also a theme in a wide swath of religions around the globe.

The Rosicrucian emerged in 17th century Europe, blending esoteric Christian teachings with Hermeticism and alchemy. They focused on understanding nature's hidden laws and developing an individual's spiritual potential.

Unfortunately, there are too many secret societies and organizations to mention in this work. While most of us have heard about the *Illuminati* and the *Order of Free Masons,* it is unlikely that many people have heard about the *Order of Gimghoul,* the *de Molay Group, Jacob's Daughters, Romeo's and Juliets,* or the *Many Worlds League.* If you're interested in learning more about mystery schools or secret societies, one of best books I ever read on the subject was *The Secret History of the World,* a New York Times bestseller by Mark Booth.

Hence, common currents run through all of them. It typically includes gradual revelations of knowledge through ritual and initiation, an integration of blended scientific and spiritual understanding, personal transformation, and the use of symbols, secret languages, and allegories to convey complex concepts.

Additionally, direct experience through frighteningly life threatening or traumatic hazing methods is highly encouraged among some secret organizations.

Let's imagine a modern mystery school from a quantum physics perspective. Such an organization let's call it, "*The Order of Ether-Nick*" (TOEN) might structure itself around the principles of quantum entanglement and the observer effect.

The practices might explore consciousness as a fundamental force that collapses quantum wave functions, influencing reality at the microscopic and macroscopic levels to thwart wars or win lotteries. This hypothetical secret society might also organize their teachings around quantum principles:

Level 1 might focus on understanding quantum superposition as a metaphor for human potential, teaching that quantum particles and human consciousness exist in multiple potential states until "observed" or actualized through intention and action.

Level 2 explores quantum entanglement as a model for human interconnectedness, developing practices and improving *mind-controlled* technology to enhance awareness of non-local connections between consciousness and matter.

Level 3 delves into the zero-point field theory, exploring plasma harnessing techniques and visualizations to interact with a fundamental field of consciousness underlying reality.

The Order of Ether-Nick might engage in such practices as group meditation to influence quantum probability fields, experiments combining advanced measurement devices with consciousness-based practices, creating coherent fields of intention, or studying observer effects and recording quantum measurements.

This secret society might operate in small, unknown *bubbles* worldwide, sharing data through quantum-encrypted communications and meeting in temporary spaces to avoid detection. Their goal could be to develop reproducible methods for consciousness to influence quantum systems in meaningful (or nefarious) ways.

Similarly, mystery schools and secret societies are but filaments in a highly interconnected field of existence. Together, they show us that science and spirituality aren't opposing forces but partners in an ongoing quest to understand and manipulate life, the cosmos, and its energy.

So next time you doodle a spiral, say a prayer, attend church, join IEEE, encounter a sacred symbol, engage with a secret society, or perform an ancient ritual, you're directly interacting with the universe's archetypal design, a cosmic electrified spark of wonder, wildness, and wisdom. Additionally, while mystery schools and secret societies did not typically employ the mathematical rigor of quantum physics or the precise circuitry of electrical engineers, they intuitively grasped some fundamental aspects of our perceived reality that science is only now *rediscovering*.

Furthermore, mystery schools and secret societies have surprising parallels with quantum physics. While these ancient traditions were not based on empirical science as we understand it today, their symbolic, metaphysical, and philosophical insights provide the frameworks for interpreting deeper mysteries of quantum mechanics. For example, many mystery schools and secret societies teach that reality is an illusion, and human perception influences the material world. Additionally, mystery schools often emphasize the oneness of all things, an idea mirrored in quantum entanglement and Bohm's interconnectedness. This suggests an underlying unity beyond time and space.

The Pythagoreans and Kabbalists discuss reality as being composed of emanations of energy. Modern physics describes matter as energy condensed into vibrational states, similar to the quantum field theory. Likewise, such secret teachings typically describe multiple dimensions of existence (physical, astral, etheric, causal, etc.), while quantum physics suggests the existence of multiple dimensions.

Ultimately, mystery traditions teach that consciousness is non-local and transcends the physical body, much like quantum non-locality. And some interpretations of quantum physics, like the holographic principle, suggest the universe may be a projection of a higher-dimensional reality, which resonates with mystical teachings about "shadow" realms.

CHAPTER TWENTY-ONE

Breaking the Simulation
Human Potential in a Holographic Reality
And Other Shadow Works

Shadow work is a process. It means, loosely, exploring and integrating the subconscious or "shadow" aspects of oneself—repressed or ignored traits, desires, emotions, and thoughts deemed unacceptable, embarrassing, or undesirable. The concept of shadow work is rooted in the research of Carl Jung, a psychologist who introduced the idea of the *shadow* as part of the psyche.

Psyops, on the other hand (or Psychological Operations), refers to mostly secret strategic activities and/or campaigns designed to influence emotions, beliefs, attitudes, and behaviors of individuals, groups, or entire populations. These operations are typically conducted by military, intelligence, or governmental organizations as part of broader efforts to achieve certain objectives. The idea is to weaken morale or otherwise influence a population to shape perception or to gain support for a specific initiative. Modern psyops is ongoing through media and social app campaigns. Bots, humans, and algorithms are now used to target certain audiences to prompt connection, an emotional reaction, or calls to action.

There isn't a human on Earth impervious to psyops—even the people who create psyops. Think about this: What is your favorite sport, music, car, color, political party, religion, type of gun, or video game? Why is it your favorite? What is it, exactly, about anything you *like* or claim to love that motivates or moves you to play or respond or give money? What about all the things you loathe? Why do you loathe them?

It's highly likely that somewhere, a successful psyops was implemented from a distance or behind the scenes to influence your thinking and actions on a particular subject, object, or group of people. This psyops might arrive in the form of effective advertising, institutions, charismatic leaders, songs, or even, unfortunately, a traumatic event.

Psyops can be both positive and negative—and cannot be completely avoided, absent total isolation, when traversing life. However, raising *awareness* of psyops helps one circumvent exploitation, question information, protect privacy, and enhance autonomy.

Understanding our shadow psyche begins with examining aspects of the self that we unconsciously deny or suppress, including anger, jealousy, fear, shame, and even hidden talents or desires. Most of our cemented programming takes place during our formal years of childhood between the ages of three to eight.

Authority figures such as parents, grandparents, social workers, counselors, schoolteachers, church elders etc. convey societal norms to children. They do this to forge productive and civil adults for the future. If little humans were left to fend for themselves as children, its likely most of us would die early through accidents or grow up to act no better than wild animals.

Yet through this same behavioral modification and conditioning, societal, cultural, familial, or personal standards manifest in unexpected ways. While we learn from others, we also inadvertently take on negative opinions, feelings, criticisms, or thoughts of others to the negation of forming our own.

The benefits of shadow work help us resolve, heal, and integrate internal conflicts and regain a sense of wholeness. It improves our interpersonal relationships because we recognize that most of what people say or do to us, really isn't about us, as much as it is where *they* are or were in *their* psyche.

Moreover, shadow work is one of the best ways to unlock hidden strengths and skill sets.

While I am a proponent of shadow work, it can be emotionally intense and uncomfortable because it involves confronting difficult truths about oneself and requires personal honesty, vulnerability, and patience.

In spiritual practices, shadow work is often seen as a path to enlightenment or transcendence. It helps individuals align with their true nature. And sometimes that true nature might appear *out-of-this-world!*

Thankfully, in science fiction, levitation, teleportation, light beings, time travel and talking with the dead are all possible. If humans are a fiction of a holographic universe, a fragment of the Source expressing itself as a wave form, might humans be capable of the same feats as their favorite sci-fi superheroes? I feel that we can and some of us do.

We already have the capability to create in our reality what was once the stuff of science fiction. For example, most of us now carry a computer in our pocket and call it a smart device. It is an all-in-one technical gizmo that has practically become a quantum human appendage. Cue the virtual reality and algorithms, mostly negative, which feed on this illusory ego.

Like it or not, we live in a simulation, at the ever-growing mercy of these algorithms and technology. This is gloriously uncomfortable to think about. Humans would rather believe we are the masters of our domains—and other people. In a 3-4D world, we are not. There is only one temporary escape...

...we must transcend the matrix and maneuver *through* space and *around* time to even come close to the substrate that makes up everything below or beneath or around a Planck length. It is the human ego that prevents us from doing so. The ego is not our master but our container—and it is an illusion.

In this world, we are led...by our desires, fear, and programming to grasp for things or react, over meditating on a response. Each mindless push we commit blinds us to the greater mysteries of the universe, one found above, all around, and within us—a comedic drama of grand supernatural proportions if there ever was one.

What's more, we don't have to look too far back in time to remember how church leaders and many philosophers (precursors to modern scientists) insisted that we once lived in a flat world where flying machines weren't possible, and illness was the result of evil curses or God's wrath.

This was a time when the Earth was the center of the universe, the sun revolved around it, human beings were forced into slavery, human sacrifice was real to appease the gods, and only elites and church leaders were permitted to read. It was a scary place where roles were tightly controlled and people who skirted those roles were subjected to grand suspicion and harsh punishment. If someone dared push back against established boundaries or class systems, "off with her head!"

We burned people (mostly women) at the stake or bound them up and threw them into bodies of water to see if they'd float, after accusing them of witchcraft. If they sank, they were innocent (and usually dead from drowning).

Our ancestors thought up and then built quite effective torture devices such as stretching machines, guillotines, thumb breakers, and other bone crushers or painful poisons. We locked people in damp dungeons or chained them to walls until they starved to death and rotted away. We sometimes buried humans in dirt or sand up to their necks and poured honey over their heads until the ants and birds picked away at their skin and eyes. We stoned people to death in almost the same way, using handheld rocks purposely aimed at the head of an immobilized but utterly terrified, typically screaming human.

Anyone brave enough to claim, *back then,* that certain plants could heal or that the earth revolved around the sun, or that we should rethink slavery, were accused as heretics, traitors, or possessed by the devil.

Much later, in 1692, seventeen women in the from Salem, Massachusetts were branded as witches and hanged. Additionally, some of our foremothers in American society, who fought for the right to vote in the early part of the twentieth century, were hosed, raped, dragged through the streets by their hair, and/or locked up in local jails.

I must ask: As a species, have we much changed?

To illustrate, in today's world, it's perfectly acceptable—even commendable—to believe in God, to pray, and to express faith. But if you say you've *seen* God, or that God has directly *spoken* to you, or you've *encountered* an angel, an interdimensional being, or an extra-terrestrial, the reaction shifts dramatically.

Suddenly, you're not faithful—you're questionable. You possibly face swift dismissal from your job, the loss of friends, and the erosion of your credibility. People tend to respond not with curiosity, but with ridicule. You're labeled delusional, ostracized, or dismissed as a fraud. On rare occasions, you're elevated—cast as a shaman, a miracle worker, a contactee, or a guru—but more often, you're savagely pushed to the margins of society.

Likewise, if someone practices a religion outside of our entrenched cradle-raised belief systems, we grow suspicious of them, if not outrightly hostile. But we typically no longer burn people or otherwise torture them—unless we have a clearance that permits such acts for the sake of national security—or are part of a closely guarded secret society.

To further illustrate, even today, if a scientist makes a mind-shattering, world-altering discovery, his or her *science* is either confiscated for national security, discredited, or intensely disavowed and scrutinized by other scientists, or panned by religious leaders as works of the devil. Or its completely ignored or marginalized, perhaps out of resentment and jealousy, like Bohm or Tesla experienced.

Returning to the smart device example from earlier, what was once the domain of science fiction or divine order, is now only a matter of moving technology directly into the brain using something about as small as an electron. Dark ops units and unacknowledged special access projects already know how to do this, do so, and have done so for decades using implants or smart devices—more on that later.

Such Big Brother-style controls—especially those enacted without informed consent—represent a profound betrayal of public trust and a disservice to humanity. Yet psychological operations are far from rare. A brief search into *MK Ultra* reveals just how deep and far-reaching clandestine experiments on unsuspecting citizens have gone—including on infants and children. I've been told that such interventions, though ethically troubling, are justified in the name of global security—and that they've averted more than one potential worldwide catastrophe. I say, if you wouldn't perform the experiment on your own child or grandchild, leave mine alone too.

Like it or not, we have put ourselves in a collective corner where we require a corporatized-industrial-intelligence-military complex. Why? Most human primates dependent upon this planet for life, generally, cannot play nicely together. And some humans are extraordinarily violent—and extract true thrills from power and other's shock, loss, and fear.

The moment we decide to completely trust one another and let down our guard, is the moment that people, somewhere on the planet, get hurt. It pains me deeply to say this. I'd rather we love and experience peace—but my power to impose serenity upon complex systems like masses of people, each espousing different cultures and/or differing levels of rage, influence, and a desire for revenge, is rather limited. I need backup in the form of *cohesion crusaders* and a compassionate Creator who thinks we're worth saving despite ourselves.

Again, we may try to *play* God, but we are woefully poor understudies and substitutes *for* God. Left to our own devices and ever-rising egos, our deception threshold, technology, and mounting chaos are sure to collide into a supernova of our undoing.

But what if we can break the matrix? If we can't *break* it (and maybe it's better that we don't), perhaps we can step outside of it for a bit. Yet to peer through to the other side, it helps to understand others who've spent a lot of time, technology and money going before us—and to look at our worldly history and declassified files—examining them a little for what they say and a lot for what they leave out.

Remote Viewing

Remote viewing emerged in the 1970s in the US as a military and intelligence program investigating psychic abilities for information gathering and espionage. The practice involves the use of trance states. This was an attempt to gather information about distant or unseen targets using extrasensory perception or *mind sensing*. The project was launched after learning that both the Chinese and Russian governments were spending a large budget on extrasensory perception techniques—with success.

Russia's Program

Mind-reading and other clairvoyant-type research began during the 1960s at St. Petersburg's Mikhail Bekhterev Institute (MBI). It primarily focused on "biological communications", telekinesis (moving objects using mind focus) and telepathy.

Nina Kulagina, a former Red Army veteran who drove a tank during WWII, and who called herself an *average* housewife, was studied for years at MBI for her psychokinetic abilities. If she'd been a fraud, Soviet powers at the time could have put her on the next train to Siberia…but they didn't.

Kulagina could allegedly move light-weight objects, produce heat on another's skin, and influence lasers. She became well-known outside of the Iron Curtain for her abilities and ignited a psychic arms race still in existence today.

While the West claimed she was a complete fraud and her abilities were nothing but Soviet propaganda, it was compelling enough for the United States and the UK (and Canada, Germany, Israel, and Australia) to launch its own programs, which lasted decades and cost millions of dollars.

Under the initiative of the Chief of the General Staff, General of the Army Mikhail Moiseyev, a new unit was formed. Unit 10003, led by Lt. General Alexei Savin, investigated psychic warfare applications and continued its research through the late 2000's at various academic institutes. The USSR publicly admitted to about ten remote viewers, but it was closer to one-hundred-three. Schools instituted government ordered programs too, to search out psychic *sensitives*. Kids showing promise were taken from their parents and sent to more remote schools (pun intended).

China's Program

The "Extraordinary Human Body Function" research program was controversial. It was a Chinese government initiative investigating claims of supernatural abilities in children, particularly their capability to read or identify objects while blindfolded (schools in the UK, the *ICU Academy*, and *Mindgroom*, still do the same thing today, with much success).

The Yunnan Qigong Research Institute (and others) led studies examining hundreds of children who could, while blindfolded, identify colors, pictures, text and characters or use other body parts to do so. The children had open minds, unclouded by doctrine or societal programming about what is and what isn't possible.

Traditional Chinese medicine concepts were also integrated, which included Qigong (cultivation of life energy), meridians beyond the traditional acupuncture system, and the concept of extraordinary functions using teyi gongneng (telepathy, psychokinesis, and clairvoyance).

The Russians seemed to emphasize physical mechanisms and electromagnetic theories while the Chinese incorporated traditional spiritual/energy concepts. Both programs were once more secretive than US efforts, providing much less documented or declassified evidence. Notably, both countries maintain covert programs in psychic, metaphysical, and quantum bio-tech phenomena, particularly around influencing large AI and complex computer systems or various energy generators.

The US Gateway and Stargate Programs

Back in the US, Stanford Research Institute (SRI) conducted initial remote viewing research in 1972. Its founders, if we might call them this, were two men, Hal Puthoff and Russell Targ—but remote viewing had much earlier roots, as we'll soon learn.

Puthoff has a storied career as an electrical engineer, professor, physicist and researcher. I'm convinced that the man has a wardrobe of secret clearances in his closet and dons at least a few every morning. Anyway, he's brilliant, and was a commissioned officer in the US Navy, assigned by the Department of Defense to a research group at Fort Meade, working on non-linear devices and optical components for high-speed computers.

Later, he joined the Hanson Laboratory of Physics at Stanford, where he co-authored a paper, *The Fundamentals of Quantum Electronics*. After teaching electrical engineering classes, he joined SRI's electronics and bioengineering laboratory where he focused on quantum physics (particularly, faster-than-light transmissions) *and* parapsychological studies.

Russell Targ was also a physicist at Stanford who joined Puthoff in remote viewing research. It is important to note that while Targ's work in parapsychology has been criticized for lacking scientific rigor, he was earlier involved in laser research, where he co-authored a 1962 paper describing the use of homodyne detection (a method of extracting information from coded oscillating signals) using laser light.

Consequently, a deep dive into Targ's work unveils that he and Puthoff were on the frontier of a myriad of major technological advancements—and secured some very interesting patents.

The CIA came calling about this time and provided funding for Project SCANATE in 1973, while the U.S. Army established Project STARGATE (aka the Cognitive Science Program, found alongside and under the Laboratories of Fundamental Research) in 1978, mostly at Fort Meade in Maryland (but likely managed elsewhere). The Defense Intelligence Agency programs were funded for almost twenty-years at a cost just shy of a million dollars annually. This lends credence to the idea that remote viewers could indeed provide *hits* far above chance.

The three main pioneer remote viewers at SRI were Patrick Harold Price, Uri Geller, and Ingo Swann. Swann claimed to be able to see auras from the age of three, after a tonsillectomy. He, as well as Puthoff, were devout scientologists, where, interestingly, most base teachings center around space, matter, and energy.

Geller was reportedly brought to the US in the 1970's by Andrija Puharich, a physician, inventor, and researcher. Puharich attended Northwestern University as part of the US Army's specialized programs to meet the demand for technical skills. He regularly interfaced with high-ranking officials from the Pentagon, the CIA, and Naval Intelligence.

Developing more than fifty US patents, Puharich endorsed Geller as a genuine psychic. While working with psychics like Geller, Puharich also met with various channelers, studied hypnosis, and took great interest in poltergeist activity and UFO's. He even briefly studied interfacing the human mind with super intelligent computers—a bit like the Russians.

Notably, his protégé, Geller, allegedly made millions divining (the art of dowsing) mineral sites for large corporations overseas and is also said to have worked for Mossad, Israel's foreign intelligence agency, considered one of the best in the world.

Edwin C. May, a nuclear physicist by training from Pittsburgh, also worked as a research scientist on Stargate, at Stanford Research Institute, later becoming its director. He pivoted to the private sector with Science Applications International Corporation from 1975 to 1985, which did work for the Department of Defense.

Over time, approximately two-hundred-fifty-nine US-based *sensitives* conducted slightly more than twenty-six-thousand supernatural sessions—before the program was dismantled and went dark—only to be resurrected, rebranded, and remain ongoing—in the private sector.

If this sort of telepathy, out-of-body experience, or telekinesis was the stuff of crackpot pseudoscience, why were so many legendary mostly strict reductionist scientists involved in its decades long research?

The public sector program officially ended in 1995 after a CIA-commissioned report by American Institutes for Research concluded that remote viewing showed absolutely no intelligence value. However, a declassified 2016 report counters that "sensitives" (another name for remote viewers) had merit—and that the program came to an end after the mysterious death of law enforcement officer Patrick Harold Price, otherwise known as "Pat".

A low-key and observant sort of man, Price had been a police commissioner and detective in Burbank, California before participating in several remote viewing projects at Stanford, including both Stargate and SCANATE. The CIA soon tapped Price's extraordinary abilities for espionage purposes. He was provided with maps and photos so that he could obtain information on facilities behind the Soviet lines—and he did.

Notably, Price sensed "important" things, especially regarding locations without much foreknowledge. For instance, he purchased several parcels of property in "Silicon Valley" years before anyone was calling it that. He did not understand the significance at the time but had a *hunch,* he said, they'd be valuable.

Once, Price was given coordinates by Puthoff to a Virginia vacation cabin only to begin describing a facility he remote-viewed that had to do with a pool table and files called "eight-ball", among others. Turns out he'd cracked the code and located an ultra- top-secret facility on the other side of the cabin! After heads rolled at the Pentagon, and scrambling was completed to ensure there was no internal leak, moles, or double-agents, the CIA directly hired Price as a contractor.

Unfortunately, Price died of a heart attack at the age of sixty-five while visiting Las Vegas. Was it foul-play? Some claim he was assassinated by Russian sensitives or maybe *offed* by a secretive US-based intelligence outfit. If Price's abilities were so profound, it's highly unlikely the US would kill him, which tends to care for, rather than destroy, its assets—unless some thought of Price's abilities as the work of the devil—but that's another book.

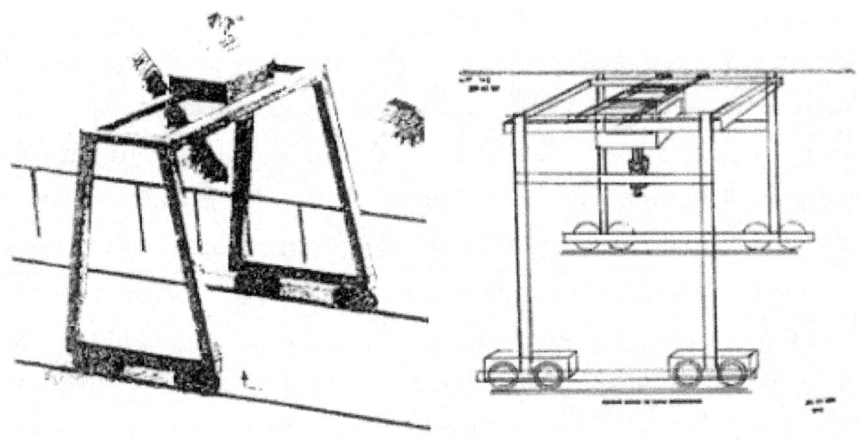

One of Pat Price's best-known sessions is the description of the Soviet arms factory in Semipalatinsk in 1974, during which he sketched and described, true to scale and in scary detail, a huge eight-wheel gantry crane and a camouflaged steel ball under construction. His drawings were later confirmed by satellite photos.

The Monroe Institute

The Monroe Institute, founded by Robert Monroe in 1971, became involved with remote viewing due to its research into altered states of consciousness. The CIA approached the Virginia based facility and its founders to explore psychic espionage potential, particularly through the development of remote viewing abilities.

US Lt. Colonel Wayne McDonell composed a research paper entitled, *Analysis and Assessment of the Gateway Project* in a bid to get INSCOM (the Army Intelligence and Security Command) to buy into the project. It provided a "lucid model of consciousness functions" for astral travel or out-of-body experiences. If it was hogwash, again, why was it endorsed as science and picked up by Army intelligence?

The paper discusses a lot about *resonance, frequency, matter, energy,* and the cosmos as a *hologram*—a bit like a quantum physics manifesto! Something called the *Absolute* is also mentioned and appears to mean unconfined energy where altered states of consciousness might gain access to alter timelines, events, or for search and rescue missions.

Thanks to a CIA operative named F. Holmes "Skip" Atwater and a former Army intelligence officer named Joseph McMoneagle, the Monroe Institute became premier training ground for teaching remote viewing. US military branches sent its intelligence officers there. In fact, it was Atwater who spearheaded the Army's remote viewing programs and a few subprograms before he took over as director of the Monroe Institute.

Monroe, I should note, was an American radio broadcasting executive. He became interested in altered states of consciousness while researching *sleep learning* and then had an out-of-body experience that initially terrified him.

Consequently, *Hemi-Sync*, short for *Hemispheric Synchronization*, was also developed by Monroe. He claimed that the technique aligns brain hemispheres, thereby creating a 'frequency-following response' or resonance. It is alleged to evoke certain effects or states of mind, where visions or clairaudience (hearing voices or music in the head) are possible.

One notable remote viewing success was Joseph McMoneagle's ability to positively identify a crashed Soviet Tu-95 bomber in Africa and its location within a mile.

Another was Operation Grill Flame, a paranormal intelligence program that unfolded, one media outlet said, like a spy thriller. Military-trained clairvoyants met more than two-hundred times to conjure up visions about the health, treatment and whereabouts of the Iranian hostages. What nobody in these clandestine circles want you to know is that the remote viewers found them. The hostages were finally released in January 1981.

Other effective remote viewers were Lynn Buchanan, part of the US Army's Stargate remote viewing program, who went on to develop training for civilian populations.

Notable female remote viewers include Hella Hammid, who demonstrated significant accuracy in SRI trials, and Frances Bryan, who successfully worked on missing persons cases.

Moreover, Angela Dellafiora Ford spent almost forty years as a highly successful intelligence analyst and was an Army remote viewing program participant at Fort Meade. The interesting thing about her is that she wasn't *discovered*. She claims in various interviews that she knew as a child she had psychic abilities. She would shut down her conscious mind to automatically write. She has located fugitives and cracked so many codes and cases for the US Army and other intelligence agencies that she became a bit of a celebrity within clandestine circles.

In addition to the Monroe Institute and Stanford Research Institute (SRI), the Mind Science Foundation (MSF) was founded in 1958 in San Antonio, Texas—decades earlier—by explorer and philanthropist Tom Slick, with the mission to investigate consciousness and extraordinary mental phenomena.

The organization conducted research into telepathy, remote viewing, and other parapsychological topics through the 1970s-80s. While less prominent than SRI, MSF helped establish early protocols for testing psychic abilities.

Before this, the most notable early remote-viewing-type experiments were carried out by Dr. J.B. Rhine at Duke University during the 1930s. These experiments aimed to test extrasensory perception (ESP) using Zener cards, which feature five distinct symbols (circle, cross, wavy lines, square, and star).

Rhine's work laid the foundation for modern parapsychology and secret military studies. Subsequent Zener card experiments were *unofficially* conducted off record or promoted by various organizations and individuals interested in ESP, psychic phenomena, or parapsychology—some even within elementary schools.

As a child, I was separated from the rest of my class for having correctly identified a bowl of plastic fruit behind a screen during a school *fun* exercise. I do not remember who the people were, but I do remember one man in a blue military uniform gazing sternly upon the class. He nodded to the teacher, and I thought I was in trouble when I was sent to another classroom.

Afterward, I was asked to guess what forms appeared on Zener cards. I was so scared; I failed miserably and didn't get a single *hit*. Toward the end of the test, I rapidly guessed so I could get out of there. A few days later, I was taken offsite to a building where I met with a dark-haired lady named Ruth. She played a board game with me called *Candyland*.

After one grueling session, where she pushed me on a guessing task, I grew frustrated and flipped the game board. She knocked on the locked door, shaken, and asked to be let out of the room (I hadn't touched the board and wondered why the door was locked). I didn't like Ruth, the *gifted* testing, or the probing questions she posed. I protested to my grandparents that I never wanted to return. I was a child unable to eloquently express that something didn't *feel* right about all of this, so I do what kids sometimes do. I turned away, shut down, and refused to speak.

My grandfather, by this time working as a subcontractor at a nuclear facility, intervened at the request of my distraught and devoutly Catholic grandmother. The sessions abruptly ended. As an adult, I've since met two individuals, with whom I later developed friendships, who relayed similar childhood experiences.

Coincidentally, successful archaeological site mapping experiments were also conducted with remote viewers, which later led to remote sensing equipment, such as ground penetrating radar and sonar transmitting through tiny pulses of high-frequency radio and sound waves.

Stephen Schwartz, a faculty consultant at Saybrook University and columnist for the journal *Explore* described, in an interview, the discovery of a previously unknown Bahamian shipwreck identified as the Brig Leander, using a team of remote viewers. But that's not all—Schwartz was the principal researcher studying the use of remote viewing in archaeology. Using the technique himself, he discovered Cleopatra's Palace, Marc Antony's Timonium, ruins of the Lighthouse of Pharaohs, and sunken ships along the California coast.

Likewise, Dr. Courtney Brown—another of the original Stargate remote viewers, started the Foresight Institute. As a promoter of non-linear mathematics for social science research, who also teaches political science at Emory University, he and his colleagues believe that remote viewing is a controlled and trainable mental process involving psychic ability. It can be used to transfer perceptual information across time and space. His nonprofit institute is dedicated to research and education in this field.

Interestingly, nonlinear mathematics is a branch dealing with systems, equations, or *phenomena* where no straight lines or relationships exist between variables. Unlike linear systems, which are relatively simple and predictable, nonlinear systems are complex. They exhibit a wide range of behaviors, including chaos, bifurcations, and emergent properties. A parabola is one mathematical example of a nonlinear equation. It is a U-shaped curve that is the graph of a quadratic equation, a fundamental geometric shape. It arises in a host of contexts, including physics, engineering, finance, and geometry.

While the topic of remote viewing remains controversial, with ongoing debate between proponents who claim statistical evidence for remote viewing abilities and skeptics who attribute results to confirmation bias, propaganda, or methodological flaws, our deep dive into quantum physics warrants further investigation of this modality.

Generalized Remote Viewing Protocols

In the event any reader would like to try remote viewing I've compiled a few basic steps gleaned from various declassified military manuals and websites. Fair warning: I'm not a fan of the military manuals on remote viewing. I feel the military and intelligence agencies purposely obfuscate the process and make it much more complicated than it need be. While I cannot personally or otherwise guarantee positive results, as this book is a work of fiction, I wish you the best of success!

1. Preparation

a. Mental and Physical Environment

- **Quiet Space:** Choose a distraction-free, quiet environment that fosters focus and minimizes external interference.
- **Comfortable Position:** Sit or lie down comfortably, ensuring good posture to prevent physical discomfort during the session.
- **Meditative State:** Begin with deep breathing or meditation to achieve a relaxed, focused state. This helps quiet the mind, heighten intuition and move into a transcendent state.

b. Materials

- **Notebook and Pen:** Prepare to record impressions, draw sketches, or take automatic notes/messages.
- **Target Coordinates or Cue:** Use a random alphanumeric code (e.g., "1234-5678") that represents the target to prevent the viewer from having prior knowledge. (*Note*:* Eventually, remote viewers realized they didn't need coordinates—but could see or otherwise discern locations, messages, or files by picking up energies or cues in the quantum field all around us.)

2. Setting the Intent

- Clearly define the purpose of the session (what is the goal?)
- Repeat a focused intention mentally, such as: *"I will perceive and record the target information accurately."*
- Take your ego out of the equation. Nothing hinders an effective remote viewer faster than arrogance or fawning for attention. My best *hits* arrive when I quell the ego and treat compliments or criticism the same…with detachment. As I've learned from direct experience, the weight of your head will swiftly knock you on your ass.

3. Entering the Viewing State

a. Sensory Awareness

- Enter a state of heightened awareness (you will *feel* this rather than *know*), focusing on all senses until letting them go to just "inter-be" and connect with energy fields.
- Allow random thoughts or sensory impressions to flow naturally without judgment or interpretation.

b. Avoid Analytical Overlay (AOL)

Acknowledge that the conscious mind will prematurely interpret impressions. Note these "AOLs" without acting on them. *Example*: If you see an image of a triangle, do not immediately conclude it's a pyramid; simply note "triangle."

Perception and Recording

- Begin by noting spontaneous impressions (e.g., colors, shapes, sounds, temperatures, textures).
- Refrain from "overthinking" or "speed guessing".

- But do write or sketch whatever comes to mind without hesitation or filtering and don't second-guess or doubt yourself. Remote viewing takes practice and is a perishable skill.

Specific Stages of Remote Viewing

Ideogram: Draw spontaneous marks representing a target's essence. Do not overthink; let the hand move naturally, as if mindfully but subconsciously doodling.

Sensory Data: Describe physical qualities of the target (e.g., "smooth, metallic, warm").

Dimensional Data: Record impressions of size, scale, and shape.

Emotional/Conceptual Data: Note any feelings or abstract ideas related to the target (e.g., "peaceful," "troubling", "serene", "historical significance").

Site Sketching: Create a drawing based on impressions, emphasizing shapes and spatial relationships (some will claim detailed drawings are best but abstract drawings seem to have the best temporal "accuracy" for beginners or those who lack natural drawing ability. A few people (like me), write out their impressions—sometimes with poetry or telling a short story or using symbology or other types of characters.

Feedback and Analysis

- **Self-Evaluation:** After the session (basic sessions are twenty-minutes for beginners but hover around ninety-minutes for more advanced RVer's) compare notes and sketches to the actual target if it is available. Also note that top-secret intelligence RV'ing used to last several hours but is now enhanced with technology.

- **Cross-Validation:** Work with a neutral observer or analyst to review the session data without introducing bias.
- **Iterative Process:** If the target remains unclear, conduct additional sessions with the same cue or related cues.

Ethical Considerations

- Always approach remote viewing with respect for privacy and ethical boundaries.
- Avoid using remote viewing for manipulative, intrusive, or unethical purposes.
- It's not a competition—but like playing a sport, you'll experience good, mediocre, or even inaccurate sessions.

Tips for Remote-Viewing Improvement

Practice Regularly: Consistency and dedication are vital for honing intuitive skills.

Stay Neutral: Avoid personal biases or expectations about the target. In a 3-D world it is almost as if we are programmed for knee-jerk judgment. In remote viewing, its best to avoid snap judgements. Admittedly, this is one of the most difficult aspects of remote viewing in the beginning.

Collaborate: Working in teams or with trained facilitators greatly improves accuracy and provides *down-to-earth* perspective. Finding trained facilitators is also not easy for members of the public to locate.

Keep a Journal: Track all dates, times, sessions, successes, and challenges to identify positive/hindrance patterns and refine techniques.

As this chapter draws to a close, imagine you're on the precipice of life-changing cosmic truth that shimmers just out of reach. Remote viewing, often dismissed as pseudoscience or worse, might instead, be a portal into a vast and super-intelligent alternate field of existence with its roots deeply embedded in quantum physics. But I must ask: what makes remote viewing resonate so deeply with experiences like astral travel, out-of-body experiences (OBEs), meditation, lucid dreaming, or even precognition?

All these *enhancement* practices share one critical element: the expansion and transcendence of human consciousness. For instance, remote viewing relies on focused intent, astral travel on projecting awareness, OBEs on disembodied perception, meditation on the stilling of the mind, lucid dreaming on navigating the subconscious and controlling dream content, and precognition on transcending linear time.

Each involves stepping beyond the conventional limits of 3-4D sensory input and venturing into realms where the usual laws of space and time dissolve. Quantum physics, space-time geometry, and clandestine research on paranormal topics, to the tune of millions of dollars, offers tantalizing hints and clues that such phenomena are much more than crazy talk or metaphysical musings.

Furthermore, the *observer effect* in quantum mechanics suggests that consciousness itself plays a role in shaping reality, while concepts like *entanglement* and *nonlocality* imply connections across vast distances, bypassing the constraints of space and time.

Could remote viewing, OBEs, and other related phenomena stem from a quantum-level entanglement of consciousness with the universe? If space-time is a pliable, plasmic, controllable electrically-based field of wave-like potential, perhaps consciousness acts as a sort of laser if condensed to the human brain, beaming information between seemingly distant points that are simultaneous and part of the whole. This might explain how some effective and accurate remote viewers describe distant locations or how precognitive dreams might foretell future events.

Similarly, psychedelics—long associated with altered states—may function as catalysts, temporarily disrupting or dissolving the tyranny of the left brain and allowing the personal subconscious part of one's mind to directly interface with the quantum electrical grid—which I always envision as a cube-shaped deep purple glowing lattice type of structure surrounded by haze.

In fact, scientists discovered in 2022 that certain fungi might be doing more than just quietly growing in dark spaces—they could be, literally, chatting it up. Andrew Adamatzky, director of the delightfully sci-fi-sounding Unconventional Computing Laboratory and Professor at the University of the West of England in Bristol (yes, his job title is almost longer than a mushroom's mycelium), proposed that mushrooms may communicate using a form of electrical signaling that *mimics language.*

His team recorded electrical activity in four species of fungi and noticed that the electrical spikes often grouped together like words in a sentence. In fact, they found patterns that could represent a vocabulary of up to fifty "words." That's right—mushrooms might be whispering secrets using pulses and sparks. Sadly, no one yet knows what they're saying.

Likewise, Carl Jung's concept of synchronicity—the meaningful alignment of seemingly unrelated events—echoes through all these *supernatural* (supernatural meaning science we don't yet understand) abilities and experiences. While logic and probability might chalk up such occurrences to coincidence, intuition hints otherwise. And so does love. A human being might claim to be a rabid materialist, but I haven't yet encountered a physicist, mathematician, or engineer who claims they've never loved—nor can they describe this intangible feeling using strict empiricism. It's a bit like developing a testable hypothesis—what is the intangible *thing* in the ether that prompts a scientific *idea* or *hunch* in the first place?

All this pondering leads me to other subjects, thoughts, and questions. Synchronicities appear amplified during meditation, lucid dreaming, and especially remote viewing, as if the act of engaging with deeper consciousness tunes us into an elevated cosmic rhythm. Are synchronicities merely chance, or are they glimpses of an underlying order—the 'hidden variables' that Einstein, and later Bohm, once sought?

This brings us to *reality augmentation*: the idea that our perception of reality can be enhanced or expanded through altered states (and covert manipulation). Psychedelics, for instance, might distort reality but it also appears to reveal layers of reality normally inaccessible to the unaltered mind. Remote viewing could be seen as their sober counterpart—an intentional, disciplined effort to access these layers without chemical intervention. In both cases, the boundaries between self and other, here and there, now and then, blur, suggesting that reality is fluid and shaped by our interaction with it.

The rational mind struggles with such ideas, clinging to logic or materialism as a comfort mechanism and for conformity's sake. Yet intuition and experience—that quiet, subjective inner knowing—guides us toward truths logic and science find impossible to fathom or measure. Remote viewing exemplifies this interplay. It requires structure and discipline (logic) but thrives on subtle impressions and flashes of insight (intuition). Together, they form a higher wire, a filament between the known and the unknown currents in the quantum wave field.

As we stand at this crossroads of science, mysticism, and human potential, one question looms large: what is real? If remote viewing and its kindred practices allow us to transcend ordinary perception, they challenge the very foundation of reality as we know it. Is reality a construct—a holographic projection, as some physicists suggest? Or is it a malleable field of infinite possibility, shaped by our consciousness and intent?

Perhaps the answer lies in embracing the paradox rather than choosing between logic and intuition, synchronicity and coincidence, or science and the sacred. For in the paradox of remote viewing, some have glimpsed the infinite and had *hits* far above chance or logic. The same can be said for faith, spirituality, and intuition too. There are things we don't know and may not ever know.

CHAPTER TWENTY-TWO

Timeless Marvels
Ancient Ingenuity and Modern Masterpieces

One way to temper our egos is to consider how little we know about a multitude of more down-to-earth mysteries. Alongside ancient superstitions, quantum physics, and modern psyops, humans or some other type of being/s completed grandiose feats of engineering around the world that still, today, cannot be explained.

The brilliance of engineers, both ancient and modern, is a testament to humanity's relentless pursuit of innovation and mastery over nature's challenges. From the architectural marvels of ancient civilizations—pyramids, aqueducts, and temples—to the cutting-edge technologies of today, neutrino detectors and particle accelerators to name only two, engineers have consistently transformed theoretical ideas into tangible realities. Engineers give shape to our world. Their ability to bridge abstract concepts through six dimensions (orthogonal being one hardly anyone mentions), with practical application, surpasses even the achievements of physicists.

Without engineers, the substantial experiments and insights of physicists would remain confined to the realm of thoughts and prayers.

Engineers are the ones who design the tools, build the machinery or facilities, appropriately understand and wire the electricity, and create the systems and hardware that allow physicists to test and expand their theories. In many ways, engineers are the unsung heroes who ensure that the elegance of physics doesn't end in equations (or explosions or tripped up timelines), but instead, becomes the foundation for positive progress and innovation.

Listed below are some remarkable feats of structural masterpieces and mechanical wonders that continue to baffle modern scientists, modern engineers, and historians the globe over, due to their precise complexity. Each was built at a time when humans allegedly lacked advanced tools, knowhow, or techniques—and in some cases, a well-defined culture.

If this is true, who or what built them and what can these ancient entities each teach us about the powerfully knowledgeable and tuned in civilizations preexisting our own? And what happened to them?

1. The Pyramids of Giza (Egypt)

- **Mystery:** The precision with which the Great Pyramid of Giza was constructed, including its alignment with cardinal points and astronomical features, is astonishing. The transportation and placement of the massive stone blocks, some weighing up to 80 tons, remain a topic of debate.

- **Hypotheses:** Theories range from ramp systems to lost ancient technologies or the use of now dried up waterways, but no definitive explanation exists.

2. Stonehenge (England)

- **Mystery:** How and why the massive bluestones, weighing up to 25 tons each, were transported over 150 miles from Wales to their current site. The purpose of Stonehenge, whether astronomical, religious, or social, also remains unclear.

- **Hypotheses:** Speculations include sleds, rollers, and waterways, but no concrete evidence has been found.

3. The Nazca Lines (Peru)

- **Mystery:** These massive geoglyphs, some spanning hundreds of feet, are best viewed from the air, raising questions about how the ancient Nazca people created such precise designs without modern tools or aerial perspectives.
- **Hypotheses:** Theories suggest they were made for astronomical, religious, or navigational purposes, but their exact purpose and construction methods are uncertain.

4. Baalbek (Lebanon)

- **Mystery:** The site features the Temple of Jupiter with some of the largest cut stones in the world, including the "Stone of the Pregnant Woman," weighing over 1,000 tons. How these stones were quarried, transported, and precisely placed remains unsolved.
- **Hypotheses:** Techniques involving sledges, pulleys, and manpower have been suggested, but they seem insufficient for such enormous weights.

5. The Antikythera Mechanism (Greece)

- **Mystery:** This ancient analog computer dated to 100 BC, give or take, was discovered in a shipwreck. It could accurately predict astronomical positions and eclipses. Its intricate gears suggest an advanced understanding of engineering and astronomy far ahead of its time.

- **Hypotheses:** While reverse-engineering has revealed its functionality, how such technology was developed in ancient Greece is still not fully understood.

6. The Kailasa Temple at Ellora (India)

- **Mystery:** This rock-cut temple was carved from a single block of basalt. The scale, precision, and intricacy of the carvings suggest techniques beyond what is believed possible at the time.
- **Hypotheses:** Theories propose a combination of chiseling and advanced planning, but the sheer speed and precision of its construction are baffling.

7. Machu Picchu and Inca Stonework (Peru)

- **Mystery:** The Inca's ability to cut and fit massive stones with such precision that no mortar was required—and the stones still hold together despite earthquakes—is incredible. How they moved these stones across rugged terrain also remains unclear.
- **Hypotheses:** Techniques involving ropes, ramps, and levers are suggested, but their ability to achieve such perfection is still debated.

8. The Longyou Caves (China)

- **Mystery:** These massive, hand-dug caves were created with astonishing symmetry and precision, yet no historical records or evidence explain their purpose or construction.
- **Hypotheses:** Suggestions range from storage to military purposes, but the effort required, and the absence of records make them enigmatic.

9. The Great Zimbabwe Ruins (Zimbabwe)

- **Mystery:** This ancient city features dry-stone walls, some over 30 feet high, constructed without mortar. The precision and scale of the walls are remarkable, as is the civilization's capacity to sustain such a project.

- **Hypotheses:** It is unclear how the builders achieved such precision without modern tools or why the city was abandoned.

10. The Hanging Gardens of Babylon (Iraq)

- **Mystery:** If they truly existed, the gardens would have required an advanced irrigation system to support lush vegetation in a desert climate. No definitive archaeological evidence has been found.
- **Hypotheses:** Some suggest they might have been a myth or built elsewhere, but their exact nature remains unsolved.

11. Sacsayhuamán (Peru)

- **Mystery:** This Inca fortress near Cusco features massive stones weighing up to 200 tons, cut and fitted together with such precision that not even a piece of paper can fit between them. No mortar was used.
- **Hypotheses:** Techniques involving stone softening or advanced tools are speculated, but there is no evidence of such technologies in the Inca era. How the stones were transported and shaped remains a mystery.

12. Ollantaytambo (Peru)

- **Mystery:** This ancient Inca site includes structures with enormous andesite stones that were transported over steep terrain and assembled with incredible precision. The stones show interlocking patterns, making the walls earthquake-resistant.
- **Hypotheses:** Theories suggest sledges or rollers, but the logistics of moving multi-ton stones over mountains are not fully explained.

13. Puma Punku (Bolivia)

- **Mystery:** Part of the Tiwanaku complex, Puma Punku features megalithic stones cut with such precision that they resemble modern machine work. The H-shaped blocks are carved with intricate grooves and perfect angles.
- **Hypotheses:** Some suggest advanced tools or knowledge of geometry, but there is no direct evidence of such technology.

14. Göbekli Tepe (Turkey)

- **Mystery:** This ancient site, dated to around 9600 BCE, predates known agricultural societies yet consists of massive stone pillars arranged in circular formations. The carvings and arrangements imply advanced planning and organizational skills.
- **Hypotheses:** The construction techniques and social structure required to build Göbekli Tepe are unknown, as it predates tools typically associated with such feats.

15. Hagar Qim and Mnajdra Temples (Malta)

- **Mystery:** These megalithic temples, built around 3600 BCE, feature massive limestone blocks, some weighing over 20 tons. The stones are fitted together with remarkable precision.
- **Hypotheses:** The methods for transporting and precisely arranging such large stones on an island remain unclear.

16. The Hypogeum of Ħal-Saflieni (Malta)

- **Mystery:** This subterranean structure was carved out of solid rock with chambers that show acoustical properties capable of amplifying sound in specific frequencies. It dates to 4000–2500 BCE.
- **Hypotheses:** The technology and purpose behind such extraordinary acoustical design remain speculative, ranging from ritual use to advanced sound knowledge and healing.

17. Nan Madol (Micronesia)

- **Mystery:** This ancient city on Pohnpei Island consists of over 100 artificial islets connected by canals. Built from basalt logs, some weighing 50 tons, the construction techniques remain a mystery given the lack of tools and machinery.
- **Hypotheses:** Theories suggest basalt was floated on rafts, but how such massive stones were precisely placed is unknown.

18. Easter Island Moai and "Platforms" (Chile)

- **Mystery:** The moai statues are famous, but the precision with which the platforms (ahu) were constructed is equally impressive. Some stones appear to have been melted or fused together to form a seamless structure.
- **Hypotheses:** The exact methods used for carving, moving, and assembling the statues and platforms are not fully understood.

19. Bimini Road (Bahamas)

- **Mystery:** This underwater rock formation appears to be a man-made road or structure. The precision and arrangement of the massive limestone blocks suggest advanced engineering for its time.
- **Hypotheses:** While some argue it is a natural formation, others suggest it could be remnants of a lost civilization like Atlantis.

20. Mycenaean Cyclopean Walls (Greece)

- **Mystery:** These ancient walls, found at sites like Tiryns and Mycenae, are made of massive stones that appear too large to have been moved with Bronze Age tools. The name "Cyclopean" comes from the belief that only giants could have built them.
- **Hypotheses:** How these stones were quarried, transported, and placed remains largely unexplained.

21. The Coral Castle (Florida, USA)

- **Mystery:** Built in the 20th century by Edward Leedskalnin, this structure consists of massive limestone blocks moved and carved with unknown methods. Leedskalnin claimed to have discovered the secrets of ancient builders.
- **Hypotheses:** Theories range from simple leverage tools to claims of magnetic or anti-gravity technology, but no definitive method has been proven.

22. The Unfinished Obelisk (Egypt)

- **Mystery:** Found in a quarry in Aswan, this massive obelisk, had it been completed, would have weighed over 1,200 tons. It shows evidence of advanced stone-cutting techniques that are still not fully understood.
- **Hypotheses:** The use of copper tools on granite raises questions about the effectiveness of ancient methods.

23. The Valcamonica Petroglyphs (Italy)

- **Mystery:** These carvings, dating back to the Iron Age, depict humanoid figures with "helmets" and objects resembling modern machinery or astronomical devices. Their purpose and the level of knowledge they represent remain unclear.
- **Hypotheses:** Interpretations range from early astronomical tools to speculative theories about ancient extraterrestrial contact.

24. Derinkuyu Underground City (Cappadocia, Turkey)

Derinkuyu is a vast, multi-level underground city capable of housing up to 20,000 people, along with their animals and supplies. It extends over 60 meters (200 feet) below the surface, with sophisticated ventilation shafts, wells, and escape routes.

The volcanic tuff (a relatively soft rock) in the region made excavation easier, but the precision and scale of the work suggest advanced planning and techniques beyond basic tools.

The underground city includes stables, chapels, kitchens, storage rooms, and even schools. The level of engineering required to create such a self-sustaining subterranean system is astonishing. Cappadocia's underground cities were often connected by miles of tunnels. These passageways were narrow and defensible, with stone doors that could seal off entire sections.

- **Hypotheses:** Most experts believe it was used to protect against invaders, particularly during the Byzantine era when Christians sought shelter from Arab raids.
- **Older Origins:** Some evidence suggests that the Hittites, or even pre-Hittite civilizations, may have begun the excavation over 4,000 years ago.

Kaymaklı Underground City

Another vast underground city, Kaymaklı, is connected to Derinkuyu through a series of tunnels, though much smaller tunnels that suggest the need for secrecy and defense. The precise construction of air shafts, which still function today, indicates advanced knowledge of engineering and airflow. The exact date of construction and the reasons for such expansive underground settlements remain speculative. Like Derinkuyu, it was likely used for protection and as a long-term refuge. Its origins might trace back to the Hittites or earlier civilizations.

The Rock-Cut Churches and Monasteries Mystery

The Göreme Open-Air Museum and other parts of Cappadocia feature rock-cut churches with frescoes dating back to the

Byzantine period (10th–12th centuries). These churches were carved directly into the soft volcanic rock, yet their detailed artwork, arches, and domes rival above-ground constructions. The challenge lies in understanding how such intricate and durable structures were carved without collapsing the surrounding rock.

Cappadocia remains a testament to human ingenuity, perseverance, and adaptability. Despite ongoing research, much about its underground cities and rock formations is left to speculation, enhancing its allure as one of the world's greatest archaeological enigmas.

25. The Rose-Red City (Petra, Jordan)

- **Mystery:** Petra's most famous structures, such as the Treasury (*Al-Khazneh*) and the Monastery (*Ad-Deir*), are intricately carved directly into sandstone cliffs. The precision, scale, and craftsmanship of these structures are extraordinary, especially considering they were created around the 1st century BCE. Questions persist about how the Nabataeans, the civilization behind Petra, achieved such architectural marvels with the tools available at the time.

- **Hypotheses:** It is believed the Nabataeans used basic tools like chisels and hammers. However, the symmetry, detail, and scale suggest advanced planning and expertise. Scaffolding or platforms carved into the rock itself may have been used to reach higher sections of the facades.

Petra's Water Management System

Petra is in a desert region, yet it supported a thriving population of tens of thousands. The Nabataeans engineered a sophisticated water management system, including aqueducts, cisterns, reservoirs, and ceramic pipelines, to collect and distribute water efficiently. The precision with which the system controlled the flow of water to prevent floods and store it for dry periods is astounding.

- **Hypotheses:** The Nabataeans likely had a deep understanding of hydraulics and geology, but how they developed this knowledge and applied it so effectively remains unclear. Their ability to locate and exploit water sources in such a harsh environment continues to amaze researchers.

Petra's Treasury (*Al-Khazneh*)

The Treasury is one of Petra's most iconic structures, believed to have been a tomb or a temple. However, its exact purpose and the techniques used to carve its detailed facade, which rises over 130 feet high, are still debated. The fact that the Nabataeans began carving from the top of the cliff downwards indicates exceptional planning and expertise.

- **Hypotheses:** The Treasury may have served multiple purposes, including religious or ceremonial functions. However, no conclusive evidence exists to determine its primary use. The tools and methods used to achieve its symmetry and intricacy remain a subject of study.

Petra's Monastery (*Ad-Deir*)

The Monastery is even larger than the Treasury, with a facade measuring over 150 feet wide and 160 feet tall. Its purpose is also unclear, as it lacks inscriptions or significant artifacts inside. The effort required to carve such a monumental structure into solid rock raises questions about the labor force and techniques.

- **Hypotheses:** It might have been used for religious or social gatherings. Some theories suggest it was dedicated to the Nabataean king Obodas I and used as a place of worship.

Petra's Tombs

Petra features hundreds of rock-cut tombs, some highly ornate and others simple. Their precise alignment and intricate facades suggest an advanced understanding of geometry and aesthetics. The purpose of some tombs, particularly the more elaborate ones, and their connection to Nabataean religion remain enigmatic.

- **Hypotheses:** These tombs likely reflect the Nabataeans' spiritual beliefs, but their full cultural significance is still not understood.

The Great Temple Complex

The Great Temple is a vast structure covering over 7,500 square meters, with columns, staircases, and a large courtyard. Its purpose—whether administrative, religious, or ceremonial—remains unclear. The architectural techniques used to build this sprawling complex and its relationship to the rest of Petra are still under investigation.

- **Hypotheses:** Some believe it served as a center for political or religious activity, but its exact role in Nabataean society is unknown.

Advanced Urban Planning

Petra's layout suggests an advanced understanding of urban planning, integrating natural features like cliffs and canyons into the city's design. The Nabataeans created pathways, staircases, and courtyards that harmonized with the rugged terrain, showcasing exceptional foresight and engineering skill. Their ability to design such a cohesive city plan without modern surveying tools continues to intrigue researchers.

The Nabataeans' Lost Knowledge

The Nabataeans demonstrated remarkable engineering, architectural, and hydrological expertise, yet much of their knowledge seems to have been lost after their decline. Unlike some ancient civilizations, the Nabataeans left few written records, leaving many aspects of their culture and technology a mystery.

- **Hypotheses:** The Nabataeans' oral traditions and the lack of written documentation may have contributed to the loss of their technical knowledge.

Ultimately, Petra stands as a testament to the ingenuity and sophistication of the Nabataeans, yet it also remains one of the world's greatest archaeological enigmas. Its intricate carvings, advanced engineering, and mysterious history continue to captivate and puzzle experts and visitors alike.

All these engineering marvels and architectural feats represent the ingenuity and resourcefulness of ancient civilizations, but they also challenge our understanding of what was possible with the alleged limited technologies, tools, and knowledge available at the time. The precision, scale, and complexity of these constructions intrigue researchers and spark debates about lost knowledge, peoples, civilizations, otherworldly visitors and technologies. Quantum physics might explain some of these mysteries.

Meanwhile, quantum mechanics remains a thrilling paradox *and* a crisis. Like ancient engineering marvels, it upends everything we think we know about logical order and history. It forces us to reconsider the nature of matter and human ability. It challenges our assumptions of reality and obliterates common sense. To ask why is to fling ourselves into other realms like philosophy, where pure logic quickly falls apart when pressed against quantum states or spirituality—and then? Assumptions, intuition, and consciousness are called in to join the wizards, monks, and swamis to assess the universe's astounding practical jokes and mysteries.

Ultimately, our logical systems, like science, evolved to describe the macroscopic world of our daily experience in a classical vacuum. When we probe *reality* at the quantum level, we find phenomena that, while mathematically describable, resists explanation through purely logical deductions from classical premises.

We don't know what lies on the other side (or over or beneath) the Planck Scale or in or under places like Gobekli Tepe. Thus far, most of us are told such things haven't and perhaps can't be solved or measured—and if it can be, it might produce a larger black hole. It would be, I'm told, like taking a sledgehammer to a sandcastle.

As for discovering the nature and scale of quantum physics, we would require a photon loaded with so much energy it would collapse space-time as we know it. This suggests to me that there is something about space-time that is perhaps, fractionated, pixeled, grainy, or otherwise not infinitely devisable. But what do I know? I'm not da Vinci, Tesla, Bohm, or Einstein. I'm a curious woman gently coaxing an extraordinarily resistant crack in Schrödinger's box.

Is it possible that this "quantum vacuum" beyond the subatomic, while considered "empty" in classical physics, is teeming with fluctuating energy fields and particle-antiparticle pairs constantly appearing and disappearing?

Is it essentially, possibly, a dynamic and active "void" of infinite potentiality, where 'split' neutrinos offer possibilities for interactions with other dimensions and particle creation? Is this where we might find God and be surprised to discover His (or Her) reflection is *us?*

CHAPTER TWENTY-THREE

Breaking Boundaries

David Bohm, Nikola Tesla, J. Robert Oppenheimer, and Leonardo da Vinci were visionary thinkers who pushed the boundaries of human understanding in their respective fields—quantum physics, engineering, nuclear science, and art/science.

All four sought to understand the fundamental nature of reality through interconnectedness of the universe, harnessing unseen energies, uncovering nuclear potential in the atom, and a deep, all-encompassing curiosity about everything from human anatomy and Hinduism to the properties of light and flight. And like Claude Shannon, each found discoveries through the lost pastime of *playing*.

These revolutionaries proposed ideas that others struggled to grasp, like Bohm's "Implicate Order" suggesting a hidden, underlying unity in the cosmos or Tesla's work on wireless energy transmission and strong belief in ether.

Oppenheimer's understanding of quantum mechanics led to powerful discoveries and poignant moral dilemmas. Da Vinci had insight to inventions and ideas that wouldn't be realized for centuries. How?

All of them had *hunches, visions,* or *dreams* that often helped solve a problem or provided strokes of genius. Additionally, each courted his share of controversy and practiced some form of spiritualism.

For example, Oppenheimer wrestled with the consequences of nuclear weapons, famously quoting the Bhagavad Gita: *"Now I am become Death, the destroyer of worlds."* Da Vinci, despite designing war machines, much preferred the pursuit of knowledge and the serenity of nature over violence and conflict. Tesla studied Buddhism and met with a swami.

Looking at each of these men and the moments of time in which they were born, we witness the political machines and elite cogs at work, greedily discouraging science by leaps and with tightly controlled religious binding.

Quantum rebel **David Bohm** (1917–1992) was also blacklisted during the McCarthy era for his suspected communist ties while working on the Manhattan Project. He refused to testify against colleagues.

Additionally, his interpretation of quantum mechanics through hidden variables challenged the academically dominant Copenhagen Interpretation (Bohr and Heisenberg) and was vehemently discouraged by some of the best minds of physics—for the wrong reasons.

Bohm's theory suggests the universe has an underlying order (Implicate Order), which many physicists have dismissed as pseudoscience. He was exiled from the US and forced to work in Brazil, Israel, and the UK due to Cold War paranoia. His ideas on quantum potential, holomovement, and consciousness were largely sidelined and ignored by mainstream science and his work overshadowed by more conventional interpretations of quantum mechanics.

Nikola Tesla (1856–1943) was an eccentric genius and his rivalry with Thomas Edison was probably the greatest technological battle of all time (AC vs. DC electricity, where Tesla won the battle but lost the political, financial, and corporate war). Tesla's idea of wireless energy transmission (Wardenclyffe Tower) threatened major energy companies, leading wealthy businessman J.P. Morgan to withdraw Tesla's funding. Tesla also claimed he'd developed "Death Ray" technology, which fueled scavenger hunts for the truth and a lot of governmental interest after his death.

In fact, it was President Donald J. Trump's uncle over at the Office of Alien Property, who ordered the FBI to seize Tesla's papers and equipment after his death. Tesla's orchestrated ostracization, by being labeled eccentric and crazy, led to alleged financial ruin despite his pioneering radio, electricity, and wireless technology inventions.

J. Robert Oppenheimer (1904–1967), known as the father of the atomic bomb led the Manhattan Project, a nuclear achievement that made him both a national hero and a symbol of moral crisis. He later opposed the development of the hydrogen bomb, leading to accusations that he was a security risk during the Red Scare.

Oppenheimer struggled with extraordinary personal guilt over the bombings of Hiroshima and Nagasaki and faced humiliation when his security clearance was stripped in a widely publicized hearing, effectively ending his influence on US nuclear policy. His scientific and military allies turned on him too, for ultimately opposing and turning his back on nuclear escalation. He severely struggled with whether humanity could manage the powers it had unleashed. He feared science had outpaced wisdom and that civilization might not survive its own inventions.

Despite Oppenheimer's pessimism, it is said he didn't feel hopeless for humanity. He believed education, diplomacy, and global thinking were humanity's best — maybe only — chances at survival: *"In some sort of crude sense which no vulgarity, no humor, no overstatement can quite extinguish, the physicists have known sin; and this is a knowledge which they cannot lose."*

Near the end of his life, Oppenheimer seemed resigned and even serene. He acknowledged the beauty and terror of the universe, which often mirrors the human mind and condition in its capacity for destruction, creation, and preservation.

Leonardo da Vinci (1452–1519) remains the mysterious polymath. His dissections of human bodies for anatomical studies were considered blasphemous by the Church and his scientific ideas conflicted with religious doctrine, leading to concerns about heresy. He was accused of homosexuality in Florence (where it was illegal at the time), which led to his arrest—but he was later cleared.

Da Vinci also faced a constant struggle for patronage and had little choice but to rely upon powerful political rulers like the Medici and Sforza families to fund his work—until his final years of life when he was cared for completely by the King of France, who encouraged da Vinci to enjoy some rest, play, and live carefree.

Amazingly, da Vinci envisioned helicopters, tanks, and advanced machines centuries before the technology to create them existed. He recorded drawings of them in his notebooks. Throughout most of his life, da Vinci had to navigate the dangerous politics of religion and politics in Renaissance Italy, while seeking artistic and scientific freedom.

In the final analysis, all four men saw the potential for science and religion to either uplift or destroy humanity. Given their obsession with energy, physics, and nature's hidden forces, they hoped for a future where humanity ventures into the cosmos and human consciousness—not for conquest, but for greater *internal* understanding. Each of these visionaries also faced significant controversies and challenges, often because their ideas, ethics, or ambitions clashed with the political, scientific, or religious forces of their time.

If these four minds were to meet in a quantum space today, they might agree that humanity stands at a defining crossroads—one where wisdom and compassion must guide science, religion, and governance alike, and where paradoxes and anomalies deserve to be explored with both curiosity and rigor.

Their collective hope would likely be that we move toward a future where knowledge serves to enlighten, not destroy, and where the cosmos is not just explored, but understood as a divine, interconnected, and interactive extension of humanity and its spirit.

Science and religion—two foundational pillars of human understanding, both deeply embedded in policy and government—are often caught in a political web, constrained by dogma, hierarchy, budgetary threats, and a zealous tendency to dismiss or deny anomalies rather than explore them. It is a steep reluctance curve against progress and high-risk behavior if scientists continue to reject strange results or are replaced by demagogues.

Sadly, hope and progress collapses when it threatens the status quos of elected officials, religious or financial institutions, or academic tenure. As I stated much earlier in this book, religion and politics require enemies to fight so each may thrive. Then, consciousness itself falls into subconscious complacency, replaced by apathy, or worse, manipulated into hostility by relentless algorithms of discredit and the cataloging of sins for control. And human beings, at their most primal, are almost rabid and territorial in their pursuit of resources and protecting fragile egos.

During such times, it is only through human experience, authenticity, intuition, love, and spiritual discernment that we might distinguish between self-delusion and overt deliberate, group-justified criminal deception. Unfortunately, the dolphins, elephants, whales, and birds won't save us or this world—but depend upon us to act with care and intelligence.

If there is one resounding message within these pages, it is the call to recognize the fundamental interconnectedness of the universe, including the connection between sentient beings, their living planet, and this galaxy, revealed through the boundless power of human imagination and ingenuity—and quantum physics.

My cherished wish is that scholars, elected officials, specialists, hotshots, luminaries, intelligence mavens, and experts—those who have spent their lives mastering or micromanaging their fields—will step beyond their often short-sighted, fortified perspectives, if only for a moment, and engage with visionary thinkers from outside their financial classes and disciplines—and connect the photons beyond an electrically neutral universe. It is also hoped that we are willing to embrace new frequencies of thought and imagination.

Yes, it takes courage to shatter the comfort of established paradigms. Yet without boldness, our greatest discoveries will remain *just beyond reach* and subject to the wildly swinging pendulum of politics and primal emotions—both fleeting and poor barometers for reaching or breaching cosmic clearance and understanding interstellar consciousness.

For too long, certain scientific and religious tenets have been treated as untouchable, their truths assumed absolute, classified, or beyond question. Yet history proves that no idea, no theory, no sacred belief, or top-secret document is beyond evolution or disclosure.

Knowledge and information are not static; they are dynamic, like consciousness, and each expands with our willingness to challenge, question, and seek deeper truths. Every individual possesses enormously powerful latent faculties—hidden potentials waiting to be awakened. This should be encouraged and developed rather than consumed and discarded.

Conversely, it is sometimes true that conflict and adversity serve as catalysts for awakening. War, destruction, and upheaval have, paradoxically, fueled some of humanity's greatest leaps in understanding. But must suffering always be the price of progress? Oppenheimer would probably think not. And I'd concur.

The same fundamental energy that drives both chaos and creation—this latent and mostly untapped piezoelectric force within us—can be harnessed with intent, resonance, and focus. Directed through reason, guided by knowledge, and tempered by ancient wisdom and compassion, it has the power to elevate human consciousness and accelerate our grasp of the universe's quantum mysteries and our self-views.

Science and spirituality are not at odds; they are two expressions of the same quest for truth, separated only by language, culture, perception, classification, and interpretation. Ultimately, everything points toward a singular reality—an intelligent interconnected universe of light, immutable and eternal. Earth does not have to be a domain of struggle and suffering unless we insist upon *perceiving* it as such, refusing to acknowledge alternative possibilities for human evolution. The nature of our reality is shaped by this collective awareness.

Our unwillingness to transcend limiting worldviews and established doctrines will probably keep us shackled to the endless and unnecessary cycles of conflict, suffering, and scarcity—at least until bio-AI takes over—and then? It's over.

Human beings have always sought higher meaning and spiritual elevation, shaping our existence through work, prayer, rituals, art, technology, and even war, all to bring order to the chaos. It is why we align stones in circles, remote view, construct vast digital networks, map the stars, build engineering marvels, split an atom, invent anti-gravity propulsion, detect and harness neutrinos, and carve symbols or monuments into history's stone. It is also the same reason we dream up dark matter and energy, write fiction, exercise aimless curiosity, live off grid, or paint ethereal landscapes.

Our relentless pursuit is not random or chaotic—it is the human instinct to authentically harmonize with our world, to make sense of our place within the cosmos to try and understand the mysteries of life. It is a superposition of universal entanglement—and if we are alone in this universe, it is us, as observers and participants, who make and keep it manifest—for better or for worse—a highly important job and God-like promotion indeed.

But I do not believe for a solid quantum nanosecond that we rank so highly on the species scale that our planet and Earth are completely alone in such a vast cosmos. We have neighbors, guides, and friends, always, in dimensions and galaxies near and far to us. If we fuck it up for ourselves, we fuck it up for them too. Earth is a phenomenal repository of DNA—and therefore, *Life*, a most curious form of matter.

As we traverse the boundaries between science and the sacred, quantum mechanics, and daily 3-4D existence, a major truth emerges: the physical and the metaphysical are not separate but intricately connected.

Universal consciousness is the electrically charged architect, the experiencer, and the observer, shaping reality as much as it is shaped by it. This realization is a gateway to traversing dimensions we have long internally and collectively sensed but failed to seriously scientifi-cally explore in the public domain. We only need to hold up a mirror and step through the portal of our reflection to realize ascendancy and meet God (however defined)—and to find ourselves.

If we can break free from the constraints of flagellating religious and rigid scientific paradigms, we may yet unveil the *hidden variables* that bind us to an electrified cosmos—an infinite, interconnected expanse, where the language of light, plasma, energy, and the sacred resonance transcends all divisions. When we align with divine consciousness, the limitless potential of humanity and the universe, stands ready for our shared awakening and evolution.

We've danced around Schrödinger's box long enough.

To transcend this illusion we call reality, we must acknowledge that it was never about skinning the cat inside the box because there is no box. Forget the box. The cat is cosmic, eternal, and curled up in the lap of Source, purring in superposition—and needs to be fed. Open any box, including Pandora's, and you might find yourself staring into a mirror—one where the cat's eyes are our own.

About the Author

B.A. Crisp is a best-selling author known for her genre-blending works that fuse quantum physics, metaphysics, ancient wisdom, and speculative futurism into unforgettable stories. Her books aren't just read—they're decoded, debated, and devoured by readers who crave the intersection of science, mystery, and human potential.

Drawing from a background that spans intelligence research and human advocacy, and studying at prestigious universities such as Oxford, Ursuline College, and the George Washington University's Graduate School of Political Management, Crisp has earned a loyal following among readers. Most appreciate her ability to challenge conventional paradigms while delivering thought-provoking, page-turning fiction and nonfiction. Whether she's unraveling the enigma of consciousness, exposing buried truths in shadowy government programs, or reimagining time as a vertical construct, Crisp writes with one foot in hard science and the other in the unknown.

When she's not writing, Crisp can be found chasing mysterious particle theories, decoding ancient symbols, or sipping black coffee while contemplating the implications of Bohm's hidden variables—or possibly all three at once.

She invites readers to stay curious, question everything, and remember that reality and personal reinvention is far more fluid and possible than it appears.

To learn more about B.A. Crisp's books please visit:

www.bacrisp.com